SpringerBriefs in Ope

Series Editor

Suresh P. Sethi, The University of Texas at Dallas, Texas, USA

SpringerBriefs present concise summaries of cutting-edge research and practical applications across a wide spectrum of fields. Featuring compact volumes of 50 to 125 pages, the series covers a range of content from professional to academic. Typical topics might include:

- A timely report of state-of-the art analytical techniques
- A bridge between new research results, as published in journal articles, and a contextual literature review
- A snapshot of a hot or emerging topic
- An in-depth case study or clinical example
- A presentation of core concepts that students must understand in order to make independent contributions

SpringerBriefs in Operations Management showcase emerging theory, empirical research, and practical application in the various areas of operations management (OM), supply chain management (SCM), germane elements of Operations Research (optimization, stochastic modeling, inventory control, etc.) and all related areas of Decision Science and Analytics as applied to the practice of OM, from a global author community. Briefs are characterized by fast, global electronic dissemination, standard publishing contracts, standardized manuscript preparation and formatting guidelines, and expedited production schedules.

Ramin Rostamkhani • Thurasamy Ramayah

Navigating Circular Supply Chains

Optimizing Performance Measurement
Through Fuzzy Methods and Quality
Techniques

 Springer

Ramin Rostamkhani ⓘ
Operations Management,
School of Management
Universiti Sains Malaysia
Minden, Penang, Malaysia

Thurasamy Ramayah ⓘ
Operations Management,
School of Management
Universiti Sains Malaysia
Minden, Penang, Malaysia

ISSN 2365-8320   ISSN 2365-8339   (electronic)
SpringerBriefs in Operations Management
ISBN 978-981-97-4703-0   ISBN 978-981-97-4704-7   (eBook)
https://doi.org/10.1007/978-981-97-4704-7

This Springer imprint is published by the registered company Springer Nature Singapore Pte Ltd.
The registered company address is: 152 Beach Road, #21-01/04 Gateway East, Singapore 189721,
Singapore

If disposing of this product, please recycle the paper.

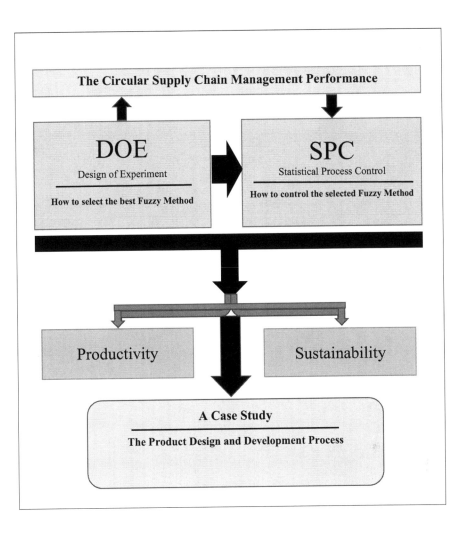

Preface

The upcoming book is one of the rarest research books on the performance measurement of circular supply chain management elements by the selected fuzzy methods through the application of famous statistical techniques, namely the design of experiments and statistical process control as the best-facilitating tools. The use of the applicable quality engineering techniques such as the design of experiments and statistical process control for the foundation of fuzzy methods in measuring the performance of circular supply chain management elements is one of the essential requirements of leading industry organizations, which provides the conditions for achieving productivity and sustainability in the industry. Although valuable books have been published about each of the concepts expressed in this book, such as statistical techniques, the performance of circular supply chain management elements, and fuzzy techniques, it is the first time that the combination of these concepts to achieve productivity and sustainability. Moreover, one of the creativities used in this book is considering the elements of supply chain management separately and combining these elements with circular thinking in the industry. Professor Thurasamy Ramayah is one of the outstanding professors in the field of information technology at Universiti Sains Malaysia with more than 30 years of teaching experience in the application of statistical techniques in production and operations management. As a senior researcher, Ramin Rostamkhani has more than 20 years of industrial experience on quality engineering techniques and elements of supply chain management. Both authors have written a book about the application of quality engineering techniques in supply chain management (SCM) in 2022. The authors of this book have exposed the results of a new approach in the performance measurement of circular supply chain management to the observation of scholars and researchers. The most important novelty of this book is the integration of traditional concepts including fuzzy methods as a main role and quality engineering techniques such as the design of experiments and statistical process control as a facilitating role in an innovative format to measure the performance of circular supply chain management that is not seen in the previous books with the same topic. It means that the combination of academic topics and industrial experience is the brilliant advantage of this new book. Moreover, this book shows how the proposed model can make an advanced background for the performance measurement of circular supply chain management on a competitive scale for the future.

The book will explain the reasons for applying fuzzy logic using quality engineering techniques in measuring the performance of circular supply chain management elements. Also, the book will tell interested readers how the design of experiments (from the quality engineering techniques collection) will select the best fuzzy method in the performance measurement of a circular supply chain management and how the statistical process control (from the quality engineering techniques collection) determines the extracted data can be under control or not. Moreover, the book explains to professional readers how the proposed model can achieve productivity and sustainability in the organization.

The authors hope to achieve a comprehensive platform by receiving the valuable opinions and suggestions of scientists and experts related to the subject of this book. Undoubtedly, any scientific research in the form of a book or article can always be modified and improved. There is no end point for being the best or becoming the best in any scientific research. Science and knowledge in every research field have no boundaries. This book is not excluded from this general rule. The authors of this book are very eager to receive comments and suggestions, especially from university professors and relevant experts and managers, to present a more complete book in the next editions. Please, do not hesitate to contact us.

This book consists of six chapters:

The first chapter contains an overview of the basic concepts used in the book. The second chapter contains the basics of the theoretical research of the book. The third chapter deals with the operational application of the book's proposed model in a sample organization. The fourth chapter describes the achievements and advantages of the proposed model, and the fifth chapter shows the future horizon in this strategic field. At the end of the book, sixth chapter will show a case study as the application of circular (lean and agility) supply chain networks in the product design and development process.

Penang, Malaysia   Ramin Rostamkhani
Penang, Malaysia   Thurasamy Ramayah

Contents

About the Authors

Ramin Rostamkhani is a Ph.D. candidate at the School of Management at Universiti Sains Malaysia (USM). He has more than 20 years of experience in quality engineering techniques. His main expertise is related to the application of quality engineering techniques and scientific approaches in supply chain management. He has recently researched circular supply chains regarding the application of fuzzy techniques as the main approach and statistical tools as the best facilitator.

Thurasamy Ramayah is a full Professor at the School of Management at Universiti Sains Malaysia (USM). He is a Professor in Technology Management whose focus of research is on technology adoption and usage in business and management using quantitative research methodology, especially the use of structural equation modeling. His mathematical approach to analyzing has had a strong impact on a better understanding of circular supply chain management dimensions.

Chapter 1
The Review of Main Concepts in the Book

1.1 Introduction

In this chapter, all the basic concepts related to the main discussion of the book are reviewed. The first basic concept is the main elements of circular supply chain management (CSCM). In the next step, after the express of performance measurement in general, a variety of common measurement methods for circular supply chain management performance are reviewed. Familiarity with the techniques of the design of experiments and the statistical process control to the extent that is required for the use of this book forms the next part of this chapter. Familiarity with the concepts of fuzzy techniques and the role impact that they can play in performance measurement forms the core of this chapter, and at the end of this chapter, the importance of achieving productivity and sustainability from the perspective of the United Nations 2030 vision document is assessed and has been emphasized. Figure 1.1 shows the main sections of this chapter.

1.2 The Circular Supply Chain Management

The best definition of circular supply chain management is the combination of circular thinking with supply chain management elements (Farooque et al. 2019). The main goal of this integration is to increase efficiency along with effectiveness (productivity) and protect the natural environment that it is surrounding us by optimizing the consumption of raw materials (sustainability). Indeed, the vital role of applying this concept (circular thinking) in supply chain management is to return the raw material to the manufacturing operations and save time and energy to reach added value. Figure 1.2 shows this integration between circular thinking and the supply chain management components (Rostamkhani and Ramayah 2022).

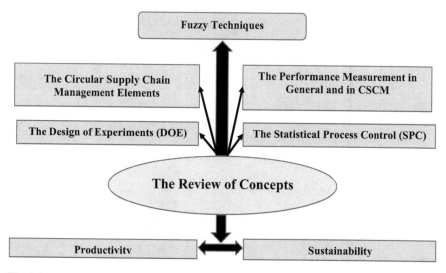

Fig. 1.1 A summary of this chapter

The Supply Chain Management Components	Circular Thinking
Customers: Determining what customers want	
Forecasting: Predicting the quantity and timing of demand	
Designing: Time and specifications that customers want	
Processing: Controlling quality and scheduling work	
Inventory: Meeting demand while managing inventory costs	
Purchasing: Evaluating suppliers and supporting operations	
Suppliers: Monitoring suppliers' quality, delivery, and relations	
Location: Determining the location of all related facilities	
Logistics: Deciding how to best move and store materials	

Circular Thinking hexagons: Redesign, Recycle, Reduce, Retrieve, Repair, Reuse, Renew

Fig. 1.2 The integration between the supply chain management elements and the circular thinking

The vital role of circular supply chain management is to create circular thinking in each component of supply chain management. This approach can reach productivity and sustainability. Circular thinking can start with the redesign and end with the retrieve. In today's competitive world, modern supply chain management entities must be willing to transition to circular supply chain management. This procedure can save time and energy related to their software and hardware resources.

1.3   The Performance Measurement (Performance Evaluation)

One of the characteristics of leading organizations is the successful use of performance measurement to find out and judge about the organization and the effectiveness and efficiency of its programs, processes, and employees. Leading organizations are not limited to collecting and analyzing performance data; these organizations use performance measurement to drive improvement and successfully translate strategies into practical actions. In other words, they use performance measurement to manage their organization. Monitoring and measuring performance and monitoring internal and environmental changes in all dimensions, accurately analyzing changes and determining their trends, comparing the organization's performance with strategic plans, and finally providing effective solutions to reduce the gap can guide organizations towards their goals. For this purpose, all organizations need a suitable performance measurement system. The performance measurement system can be defined as follows:

The performance measurement system is a systematic method of evaluating inputs (raw materials, equipment, facilities, employees, etc.), outputs (final items), conversion, processes, and productivity in a production or non-production operation based on the environmental conditions and the organization's strategy.

For several decades, the public interest in the field of performance measurement and the use of its results has been focused on the budgeting process and the allocation of financial resources of the organization and focused on financial indicators, but since about a decade ago, the inefficiency of this approach has shown the necessity of developing systems for performance measurement. A building that has a dynamic and vital relationship with the planning process of the organization. In this way, there has been a change in public attitude from performance measurement to performance management. This change of attitude has turned paying attention to the following points into key factors in the success of performance measurement and management systems:

- Performance measurement and management system needs a conceptual framework
- Effective internal and external communication is the key to the success of performance measurement
- The auditability of the results should be clearly provided and well understood
- Salary, rewards, and incentives should be related to performance measurement
- Performance measurement systems should be positive, not punitive
- Performance measurement results and progress should be shared openly among employees, customers, and other stakeholders
- In addition to data interpretation, the performance measurement system should bring knowledge to decision-makers

•**Evaluation at the individual level**

•(Evaluation of the performance of the individual's behavioral characteristics and work process)

•

•**Evaluation at the unit level**

•(Evaluation of the work process of each unit and the work results of each unit)

•

•**Evaluation at the organization level**

•(Evaluation of the organization's performance)

•

Fig. 1.3 The dimensions of performance evaluation

1.3.1   The Dimensions of Performance Evaluation

Figure 1.3 shows the dimensions of performance evaluation in general.

1.4   An Innovative Approach to Performance Measurement in CSCM

The best definition of performance measurement is a systematic approach to collecting, analyzing, and evaluating a project or plan to achieve its desired outcomes, goals, and objectives. The current methods of performance measurement in circular supply chain management are generally based on various objectives related to circularity, economic, environmental, and social performance. Moreover, the current methods are a combination of quantitative tools and qualitative studies in circular thinking. These methods want to develop the performance measurement systems in circular supply chain management that we can show the main domain of their applications in circular supply chain management.

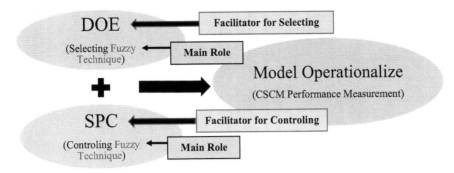

Fig. 1.4 The main construction of the book's model

In this book, contrary to all previous research that will be reviewed in the next chapter, for the first time, fuzzy techniques will be introduced in a new mixed approach to analyze circular supply chain management performance. Although there are some valuable studies about the application of fuzzy techniques in supply chain management, there is no research to combine the strongest quality techniques such as the design of experiments along with the statistical process control and fuzzy techniques in assessing circular supply chain management performance. The innovative model of this book in measuring circular supply chain management performance by identifying the best fuzzy technique utilizing the design of experiments and how to use the selected fuzzy technique with the statistical process control. This mixed approach makes a suitable possible to operationalize circular thinking in supply chain performance measurement in organizations easily. Figure 1.4 shows this innovative model in the book.

The very important issue in the model presented in this book is a new approach that has been adopted by combining statistical techniques as facilitating tools and fuzzy techniques as the main tools for measuring supply chain performance. Another significant issue is the difference between the design of experiments and statistical process control. The design of experiments is applied before identifying the suitable fuzzy technique for performance measurement, while the statistical process control is applied after the identification of the appropriate fuzzy technique (in implementing model).

Also, we will show that this mixed approach can realize increased productivity and sustainability better than any model. We should never forget the ultimate goal of this kind of research, which is to achieve productivity and sustainability at the highest possible level. The biggest distinguishing feature of this book's model is the use of a combined approach of quality engineering statistical techniques with fuzzy techniques in the performance measurement of circular supply chain management.

1.5   The Familiarity with the Design of Experiments

The unique capability of this technique in choosing the best methods or procedures to carry out any research or academic project is not hidden from any researcher. The authors of this book have proven this issue in their previous research works (Rostamkhani and Karbasian 2020; Rostamkhani and Ramayah 2022). Design of experiments (DOE) is defined as a branch of applied statistics that deals with planning, conducting, analyzing, and interpreting controlled tests to evaluate the factors that control the value of a parameter or group of parameters. This technique can determine the individual and interactive effects of various factors that can influence the output results of the relevant measurements. Also, this technique can gain knowledge and estimate the best operating conditions of a system, process, or product. Figure 1.5 is the flowchart of DOE as follows:

Figure 1.6 shows the main matrix of DOE.

The related formulas for the numerical application are as follows:

$$SS_{\text{treatment}} = \frac{1}{b} \sum_{i=1}^{a} Y_{io}{}^2 - \frac{\left(\sum_{i=1}^{a} Y_{io}\right)^2}{a \times b} \tag{1.1}$$

In this formula, $SS_{\text{treatment}}$ is the sum of treatment data, Y_{io} is the total sum related to (i) row, (a) is the number of treatment states or rows, and (b) is the number of disorder states or columns.

$$MS_{\text{treatment}} = \frac{SS_{\text{treatment}}}{a-1} \tag{1.2}$$

In this formula, $MS_{\text{treatment}}$ is the average sum related to the treatment data.

$$SS_{\text{block}} = \frac{1}{a} \sum_{j=1}^{b} Y_{oj}{}^2 - \frac{\left(\sum_{j=1}^{b} Y_{oj}\right)^2}{a \times b} \tag{1.3}$$

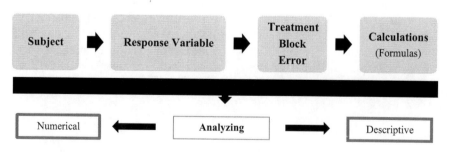

Fig. 1.5 The flowchart of DOE

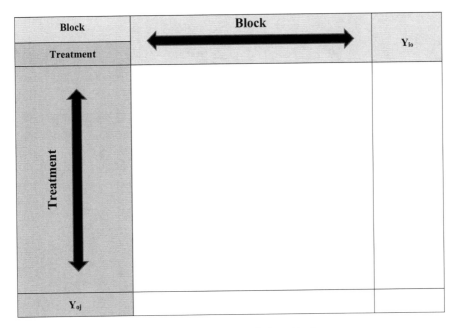

Fig. 1.6 The main matrix of DOE with one factor analysis as block

In this formula, SS_{block} is the sum of disorder data, Y_{oj} is the total sum related to (j) column, (a) is the number of treatment states or rows, and (b) is the number of disorder states or columns.

$$MS_{block} = \frac{SS_{block}}{b-1} \tag{1.4}$$

In this formula, MS_{block} is the average sum related to disorder data.

$$SS_{error} = \sum_{i=1}^{a} \sum_{j=1}^{b} Y_{ij}^2 - \frac{\left(\sum_{i=1}^{a} \sum_{j=1}^{b} Y_{ij} \right)^2}{a \times b} \tag{1.5}$$

In this formula, SS_{error} is the sum of error data, Y_{ij} is the total sum related to (i) row and (j) column, (a) is the number of treatment states or rows, and (b) is the number of disorder states or columns.

$$MS_{error} = \frac{SS_{error}}{(a-1)(b-1)} \tag{1.6}$$

In this formula, MS_{error} is the average sum related to the error data.

The main conclusion formulas are as follows:

$$F_{\text{treatment}} = \frac{MS_{\text{treatment}}}{MS_{\text{error}}} \tag{1.7}$$

$$F_{\text{block}} = \frac{MS_{\text{block}}}{MS_{\text{error}}} \tag{1.8}$$

In these formulas, $F_{\text{treatment}}$ is the statistical distribution function for treatment and F_{block} is the statistical distribution function for disturbance.

Now, if an interpretative framework is considered for the latter issue, we take a look at the following pattern:

$$\begin{cases} H_{.}: F_T \leq F_{\alpha, a-1,(a-1)(b-1)} & \textit{The treatment source does not make any significant difference} \\ H_1: F_T > F_{\alpha, a-1,(a-1)(b-1)} & \textit{The treatment source makes a significant difference} \end{cases}$$

$$\begin{cases} H_{.}: F_B \leq F_{\alpha, b-1,(a-1)(b-1)} & \textit{The disorder source does not make any significant difference} \\ H_1: F_B > F_{\alpha, b-1,(a-1)(b-1)} & \textit{The disorder source makes a significant difference} \end{cases}$$

1.6   The Familiarity with the Statistical Process Control

Statistical process control (SPC) charts—a graphic representation of the data—are drawn from the samples gathered periodically from a process and displayed on the graph in the time-ordered collected. The control limits in these charts show the intrinsic variability of a process in a stable state as the role of control charts is to help to assess the stability of a process carried out by examining punctuated data relative to the control limits.

The unique ability of this technique is to ensure the data is under control and can reach the correct results. The authors of this book have proven this issue in their previous research works (Rostamkhani and Karbasian 2020; Rostamkhani and Ramayah 2022). Another important feature of this technique is the ability to combine with fuzzy methods, which means that it can check and control the output data of fuzzy methods.

One of the most important statistical process control tools in implementing fuzzy techniques is the Z-MR control charts. They can monitor the mean and variation of data when relatively few data are considered for each fuzzy technique as in short-run processes.

Z-MR Chart estimates the mean for each character separately. It means that Z-MR Chart pools all the data for a special character and obtains the average of the pooled data. The result is the estimate of μ for that character. The character data define the groupings for estimating the process means. When we use the relative-to-size option for estimating σ, the means are also taken on the natural log of the data.

1.6.1 Z-MR Chart

- Center line

The center line represents the process average. For the Z-MR chart, the center line is always located at 0 because the data are standardized. The main formula related to Z is as follows:

$$Z_i = \frac{X_i - \mu_i}{\sigma} \tag{1.9}$$

- Lower control limit (LCL)

$$LCL = X - L \times \sigma \tag{1.10}$$

The lower control limit is $-L \times \sigma$. It is always -3 in the Z-MR chart because the data are standardized.

- Upper control limit (UCL)

$$UCL = X + L \times \sigma \tag{1.11}$$

The upper control limit is $L \times \sigma$. It is always 3 in the Z-MR chart because the data are standardized.

The relevant terms have been explained as follows:

Term	Description
X_i	Observation i
μ_i	Mean for the group
σ	Standard deviation for the group
w	Width of moving range

1.7 The Review of Popular Fuzzy Techniques in a General Perspective

Before entering the topic of reviewing the most famous fuzzy techniques, we list the reasons for using fuzzy logic and its popularity in this research.

- Fuzzy logic is easy to understand from a conceptual point of view. The mathematical concepts of fuzzy reasoning are acceptable. What made the fuzzy method is the "naturalness" of the relevant approach and not the unattainable complexity.
- Fuzzy logic is flexible and it is easy to add more functionality to any given system without starting from scratch. Fuzzy logic is resistant to imprecise data. If you look carefully around you, you will see that everything around us is imprecise. After all, most things are imprecise even under close inspection.
- Fuzzy reasoning builds this understanding into the process rather than taking it out of it. Fuzzy logic can model non-linear functions with any degree of complexity. You can create a fuzzy system to match any set of input–output data.
- Fuzzy logic can be made based on the experience of experts. Unlike neural networks, which take training data to produce impenetrable fuzzy models, fuzzy logic allows you to rely on the experiences of people who already understand your system.
- Fuzzy logic can be combined with conventional quality engineering techniques. Fuzzy systems do not necessarily replace conventional control methods. In many cases, fuzzy systems complement and strengthen them and make their implementation easier.
- Fuzzy logic is based on natural language. The basis of fuzzy logic is derived from human communication. This observation underpins most other statements about fuzzy logic.

Fuzzy methods can help researchers achieve the most suitable results more than any other approach in precise scientific topics that are formed based on imprecise data. The realization of this fact is clearly visible in the performance measurement of a circular supply chain management. In the combined case where several fuzzy techniques are used, usually the first fuzzy technique directly uses the fuzzy concept and the last fuzzy technique determines the degree of importance and influence of the outputs of the previous fuzzy techniques.

1.7.1 Fuzzy Analytical Network Process

1.7.1.1 ANP and FANP Concepts

Analytic network process (ANP) is a mathematical theory, developed by Thomas L. Saaty in 2008, to identify decision-making priorities of multiple variables without establishing one-way hierarchical relationship among decision levels. The fuzzy analytic network process (FANP) method is a widely used multi-criteria to handle interaction among the criteria and linguistic variables.

1.7.1.2   FANP Matrix

The matrix of network analysis process can be described in four main steps:

- Constructing a problem network
- Pairwise comparisons and large matrix formation
- How to calculate in the network analysis process
- Constructing a large matrix in the process of network analysis

In general, there are the normal type and its fuzzy type. Because human judgments and priorities are mostly ambiguous and complicated, decision-makers cannot express their priorities with a precise scale. Fuzzy approach and use of language variables 0–9 can be used in these cases. To complete the matrix of paired comparisons, the decision-maker is asked to make a set of comparisons as follows (Table 1.1):

The structure of the matrix of fuzzy pairwise comparisons is generally according to formula (1.12).

$$\tilde{A} = \begin{pmatrix} 1 & \tilde{a}_{12}^{\alpha} & \cdots & \cdots & \tilde{a}_{1n}^{\alpha} \\ \tilde{a}_{21}^{\alpha} & 1 & \cdots & \cdots & \tilde{a}_{2n}^{\alpha} \\ \vdots & \vdots & \vdots & \vdots & \vdots \\ \tilde{a}_{1n}^{\alpha} & \tilde{a}_{n2}^{\alpha} & \cdots & \cdots & 1 \end{pmatrix} \tag{1.12}$$

In this matrix, α represents the confidence level and μ is the optimization index, which is determined by the decision-maker. Any fuzzy number can be converted into an interval and then into a non-fuzzy number (Crisp). Experts perform a series of pairwise comparisons in the form of influence and direction within the necessary criteria, the results of which are shown in an $n \times n$ matrix such as (1). $\tilde{a}_{ij}$ is the degree of influence that criterion i has on criterion j.

$$\tilde{a}_{ij} = \left(l_{ij}, m_{ij}, u_{ij} \right) \tag{1.13}$$

$$\tilde{M}_{\alpha} = [l^{\alpha}, u^{\alpha}] = [(m - l)\alpha + l, -(u - m)\alpha + u] \forall \alpha \in [0, 1] \tag{1.14}$$

$$\tilde{a}_{ij}^{\alpha} = \mu a_{iju}^{\alpha} + (1 - \mu) a_{ijl}^{\alpha} \forall \mu \in [0, 1] \tag{1.15}$$

After completing this matrix of fuzzy comparisons and performing the necessary calculations using the above two relations, a special vector should be used to determine the priority of each criterion or sub-criteria.

	Values	Fuzzy numbers
Table 1.1 The values and fuzzy numbers in FANP	(1,1,3)	˜1
	(1,3,5)	˜3
	(3,5,7)	˜5
	(5,7,9)	˜7
	(7,9,11)	˜9

$$AW = \lambda_{\max} w \tag{1.16}$$

The weighted supermatrix W is multiplied by a certain number to obtain the limiting supermatrix (weighted limiting supermatrix) and $\lambda_{\max}$ is the mean of compatibility vector elements. In other words, the weighted supermatrix reaches the g-power to converge and obtain a stable supermatrix for calculating the general effect-priority vectors, which are called ANP weights.

$$W^* = \lim_{g \to \infty} (W_w)^g \tag{1.17}$$

The ANP weights for each criterion can be calculated using the above formula. In this formula, g can represent any number as a power.

1.7.2   Fuzzy Decision-Making Trial and Evaluation Laboratory

1.7.2.1   DEMATEL and FDEMATEL Concepts

Decision-making trial and evaluation laboratory (DEMATEL) is considered as an effective method for the identification of cause–effect chain data of a complex system. It deals with evaluating interdependent relationships among factors and finding the critical ones through a visual structural model. The fuzzy decision-making trial and evaluation laboratory (FDEMATEL) method is used to assess causal relations of accidents for construction processes. This combination is used for the imprecise and subjective nature of human judgments. Interval sets are used rather than real numbers in fuzzy set theory. Linguistics terms are converted to fuzzy numbers.

1.7.2.2   FDEMATEL Matrix

In the fuzzy DEMATEL, a fuzzy comparison matrix can be used. This fuzzy method defines fuzzy number and is more straightforward and easier to use for decision-makers. In multi-criteria problems, we have m options $A_1, A_2, \ldots, A_m$ and we want to choose the right option among these options according to n criteria $C_1, C_2, \ldots, C_n$. In addition, we also have k decision-makers who determine the degree of importance of the criteria compared to the alternatives, as well as the weight of the criteria. Multiple attribute decision making (MADM) problems are usually considered as a decision matrix. You can see a total frame of this matrix in Eq. (1.18).

Table 1.2 The values and fuzzy numbers in FDEMATEL

Values	Fuzzy numbers
(0,0,1)	~0
(0,1,2)	~1
(1,2,3)	~2
(2,3,4)	~3
(3,4,5)	~4

$$I = \begin{matrix} C_1 & & C_n \\ \begin{pmatrix} \tilde{X}_{11} & \cdots & \tilde{X}_{1n} \\ \vdots & & \vdots \\ \tilde{X}_{m1} & \cdots & \tilde{X}_{mn} \end{pmatrix} \end{matrix} \tag{1.18}$$

where x_{ij}^k is the degree of importance given by the k_{th} decision-maker for the option A_i relative to the criterion C_j. W_{jt} is the weight of the j_{th} criterion assigned by the t_{th} decision-maker. As you can see, the degree of importance of the criteria compared to the options and the weight of each criterion is expressed as a fuzzy number. To obtain the degree of importance and overall weight, we can calculate based on Eqs. (1.19)–(1.21) as follows:

$$w_{jt} = (a_{jt}, b_{jt}, c_{jt}) \qquad j = 1, 2, \ldots, n \qquad t = 1, 2, \ldots, k \tag{1.19}$$

$$\tilde{x}_{ij} = \frac{1}{k} \left(x_{ij}^{1} + x_{ij}^{2} + \ldots + x_{ij}^{c} \right) \tag{1.20}$$

$$\tilde{w}_{it} = \frac{1}{k} \left(w_{it}^{1} + w_{it}^{2} + \ldots + w_{it}^{c} \right) \tag{1.21}$$

At this stage, the development of relationships within and between indices should be done using the opinion of experts through paired comparative analysis. First, to measure communication in the form of fuzzy numbers, it is necessary to define a comparison scale. The values and fuzzy numbers have been defined as follows (Table 1.2):

The experts made a series of pairwise comparisons in the form of influence and give its results in a matrix $\tilde{A}$ to perform within the necessary criteria. The required definitions are as follows:

$$\tilde{a}_{ij} = (l_{ij}, m_{ij}, u_{ij}), \quad s = \frac{1}{\text{Max} \sum\limits_{i,j=1}^{n} u_{ij}}, \quad \tilde{X} = s \times \tilde{A} \tag{1.22}$$

The fuzzy total correlation matrix is defined as follows:

$$\tilde{T} = \frac{\tilde{X}}{\left(I - \tilde{X}\right)}, \quad \tilde{t} = \left(l'_{ij}, m'_{ij}, u'_{ij}\right)$$

$$\text{Matrix } \left[\tilde{l}_{ij}\right] = \frac{\tilde{X}_l}{\left(I - \tilde{X}_l\right)}, \quad \text{Matrix } \left[\tilde{m}_{ij}\right] = \frac{\tilde{X}_m}{\left(I - \tilde{X}_m\right)}, \quad \text{Matrix } \left[\tilde{u}_{ij}\right]$$

$$= \frac{\tilde{X}_u}{\left(I - \tilde{X}_u\right)} \tag{1.23}$$

1.7.3   Mixed Mode (The Combination of FDEMATEL and FANP)

This mode is determined by considering the total effect matrix and the opinion of experts or decision-makers. The new matrices have been constructed as follows (Eqs. 1.24–1.26):

$$T_\alpha = \begin{bmatrix} t_{11}^\alpha & \cdots & t_{1j}^\alpha & \cdots & t_{1n}^\alpha \\ \vdots & & \vdots & & \vdots \\ t_{i1}^\alpha & \cdots & t_{ij}^\alpha & \cdots & t_{in}^\alpha \\ \vdots & & \vdots & & \vdots \\ t_{n1}^\alpha & \cdots & t_{nj}^\alpha & \cdots & t_{nn}^\alpha \end{bmatrix} \tag{1.24}$$

$$T_s = \begin{bmatrix} t_{11}^\alpha/d_1 & \cdots & t_{1j}^\alpha/d_1 & \cdots & t_{1n}^\alpha/d_1 \\ \vdots & & \vdots & & \vdots \\ t_{i1}^\alpha/d_i & \cdots & t_{ij}^\alpha/d_i & \cdots & t_{in}^\alpha/d_i \\ \vdots & & \vdots & & \vdots \\ t_{n1}^\alpha/d_n & \cdots & t_{nj}^\alpha/d_n & \cdots & t_{nn}^\alpha/d_n \end{bmatrix} = \begin{bmatrix} t_{11}^s & \cdots & t_{1j}^s & \cdots & t_{1n}^s \\ \vdots & & \vdots & & \vdots \\ t_{i1}^s & \cdots & t_{ij}^s & \cdots & t_{in}^s \\ \vdots & & \vdots & & \vdots \\ t_{n1}^s & \cdots & t_{nj}^s & \cdots & t_{nn}^s \end{bmatrix} \tag{1.25}$$

$$W_w = \begin{bmatrix} t_{11}^s \times W_{11} & t_{21}^s \times W_{12} & \cdots & \cdots & t_{n1}^s \times W_{1n} \\ t_{12}^s \times W_{21} & t_{21}^s \times W_{22} & \vdots & & \vdots \\ \vdots & \vdots & \cdots & \cdots & t_{ni}^s \times W_{in} \\ \vdots & \vdots & \cdots & \cdots & \vdots \\ t_{1n}^s \times W_{n1} & t_{2n}^s \times W_{n2} & \cdots & \cdots & t_{nn}^s \times W_{nn} \end{bmatrix} \tag{1.26}$$

T_α is the main matrix, T_s is the normalized matrix, W is the unbalanced super matrix, and W_w is the balanced super matrix. In the combined mode where the FANP is combined with other fuzzy techniques such as FDEMATEL, the outputs of other fuzzy techniques can be used to determine the degree of importance and degree of influence using the FANP.

1.8   The Importance of Achieving Two Goals: Productivity and Sustainability

Productivity and sustainability benefit the economic growth of each country, as sustainable industrialization is responsible for lifting communities out of poverty. With this research, we aim to show the role of productivity and sustainability to achieve the eighth and ninth goals of the 2030 UN Agenda by creating the development for today's organizations.

Productivity: The concept of productivity is commonly defined as a ratio between the output volume and the volume of inputs. In other words, it measures how efficiently production inputs, such as labor and capital, are being used in an economy to produce a given level of output. Also, productivity is considered the combination of effectiveness and efficiency in organizational processes.

Sustainability: The concept of sustainability is composed of three pillars: economic, environmental, and social also known informally as profits, planet, and people.

SDGs 8: The eighth goal of the 2030 UN Agenda aims to achieve higher levels of economic productivity through diversification, technological upgrading, and innovation, including through a focus on high-value added and labor-intensive sectors (Productivity). Also, this goal can create the sustained and inclusive economic growth and can drive progress, make decent jobs for all, and improve living standards (Sustainability).

SDGs 9: The ninth goal of the 2030 UN Agenda aims to build resilient infrastructure, promote inclusive and sustainable industrialization, and foster innovation. Therefore, the objectives must generate employment and income in the best conditions (Productivity). Also, they must bring prosperity and build sustainable and prosperous societies around the world (Sustainability).

1.8.1   The Role of Productivity and Sustainability in the Developed Organizations

Productivity and Sustainability can help all organizations to achieve continuous improvement. Figure 1.7 shows the role of productivity and sustainability in continuous improvement.

A very vital issue when studying all contents related to productivity and sustainability is the combination of these concepts in different studies. For example, in some research, they look for sustainable productivity, and in others, they study sustainability including productivity as a sub-index in many cases. In this book,

Fig. 1.7 The role of productivity and sustainability in continuous improvement

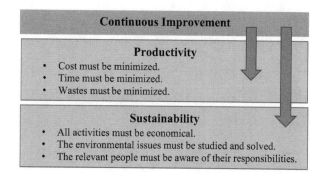

these two concepts are considered at the same level. Figure 1.7 has introduced three main indices for both productivity and sustainability.

1.8.2 The Role of Productivity and Sustainability in Creating Added Value

Added value can be defined as the difference between a particular products' or service's final selling price and the direct and indirect input used in making that particular product or service. It can apply to products, services, companies, management, and other areas of business.

Added value gives customers incentives to make purchases and increases a company's bottom line. There are many ways that companies can find a competitive advantage and bring perceived added value to their products or services including:

Productivity
- Providing features that make the product or service to be low cost with a short-delivered time and high quality. The simultaneous realization of these three features is very important.
- Making a situation for eliminating the weakness of products or services (decreased waste).

Sustainability
1. Producing a new product or service in an economical format.
2. Offering a new product or service in a recyclable format.
3. Responsibility training to producers and consumers.

1.9 Conclusion

In this chapter, we reviewed all the concepts needed to understand the main topic of this book and found the necessary preparation for the next chapters. The main line of the path in the research conducted in this book is based on the following four key issues:

1. The elements of circular supply chain management.
2. Selected statistical techniques as the strong facilitating tools.
3. Well-known used fuzzy techniques for analyzing the performance of circular supply chain management.
4. Achieving productivity and sustainability and having indicators to manage them before and after the implementation of the book's model.

References

Farooque, M., Zhang, A., Thürer, M., Qu, T., & Huisingh, D. (2019). Circular supply chain management: a definition and structured literature review. *Journal of Cleaner Production, 228*, 882–900.

Rostamkhani, R., & Karbasian, M. (2020). *Quality Engineering Techniques: An Innovative and Creative Process Model* (1st ed.). Taylor and Francis Group/CRC Press. https://doi.org/10.1201/9781003042037

Rostamkhani, R., & Ramayah, T. (2022). *A Quality Engineering Techniques Approach to Supply Chain Management* (1st ed.). Springer Nature. https://doi.org/10.1007/978-981-19-6837-2

Chapter 2
The Framework of the Research Structure

2.1 Introduction

In this chapter, the logical process in research structure is included to maintain the academic framework of this book. By reading this chapter, the readers of the book, especially graduate students and professors, will get to know what we want to achieve in this book. Although the content of this chapter and the arrangement of its contents are very similar to academic theses, it seems necessary for a better understanding of the following chapters. Please pay attention to the sequence of contents of this chapter carefully. Figure 2.1 shows a summary of the contents of this chapter.

2.2 Research Problem in the Book

The subject of research in this book is to focus on four key issues in order to achieve productivity and sustainability in the industry. These key topics are:

1. The elements of circular supply chain management.
2. Statistical techniques such as the design of experiments and statistical process control.
3. Well-known fuzzy techniques in the performance measurement analysis of circular supply chains.
4. The indicators of managing productivity and sustainability before and after implementing the model.

These key issues have been studied in a scattered manner in the previous research of scientists and related researchers, which will be considered in the background of the research, but in this book, for the first time, they have been considered in an integrated format to achieve productivity and sustainability. In addition, the

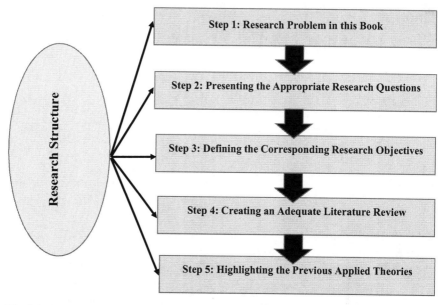

Fig. 2.1 A summary of this chapter

combination of circular thinking in each element of supply chain management is considered a new approach in this category.

2.3   Presenting the Appropriate Research Questions

The research questions always determine the direction of the research and guide the reader to find answers related to the research topic. The appropriate research questions are as follows:

1. What are the elements of circular thinking in supply chain management?
2. What are the suitable quality engineering techniques that can be utilized as a facilitation role in analyzing the circular supply chain management performance?
3. What are the best and most practical fuzzy techniques that can be selected in the main analysis of circular supply chain management performance?
4. How can the proposed statistical techniques facilitate the implementation of fuzzy techniques to analyze the performance measurement of the elements of circular supply chain management?
5. How are the topics of sustainability and productivity realized through the implemented model conducted in this book? What is the relationship between the performance measurement of circular supply chain management and productivity and sustainability? Does it exist any effective indicator to show the management of productivity and sustainability before and after implementing the book's model?

2.4   Defining the Corresponding Research Objectives

The research objectives always determine the destination of the research and guide the reader to reach the desired goals. The corresponding research objectives are as follows:

1. Proposing a new approach to circular supply chain management by integrating circular thinking into all elements of supply chain management.
2. Proposing the facilitation role of at least two statistical techniques including the design of experiments and statistical process control in the process of main analyzing fuzzy techniques for the performance measurement of circular supply chain management elements.
3. Proposing an advanced approach in using the selected fuzzy technique for measuring the performance of circular supply chain management elements in the form of required matrices and determining the relevant priorities. This goal is the main part of the research done in this book.
4. Achieving productivity and sustainability by measuring the performance of circular supply chain management elements with fuzzy techniques and also introducing innovative indicators to manage productivity and sustainability processes before and after the implementation of the model.
5. Achieving an advanced vision in measuring the performance of circular supply chain management elements by introducing new techniques and differential equations.

2.5   The Main Concepts of the Theoretical Research Structure

Since this unique research seeks to find a suitable combination of fuzzy methods and quality engineering techniques in the elements of circular supply chain management at the macro level, it requires data that must be extracted at the international level to obtain a comprehensive model. It is related to the expert's views on the relevant matrices in circular thinking in each element of supply chain management. A feature of fuzzy logic which is of particular importance to the management of uncertainty in expert systems is that it provides a systematic framework for dealing with fuzzy quantifiers. Nevertheless, to prove the achievement of productivity and sustainability as a result of the implementation of the model, the authors try to have a comprehensive approach in the used research methodology. The research methodology will be based on defining the main criteria and data analysis methods.

2.5.1   Main Criteria for the Data Quality of the Appropriate Model

The five main criteria used to measure data quality to design a model for implementing the book's model in the circular supply chain management elements are as follows:

- *Accuracy:* Whatever data is described; it needs to be accurate.
- *Relevancy:* The data should meet the requirements for the intended use.
- *Completeness:* The data should not have missing values or miss data records.
- *Consistency:* The data should have the required strength and can be cross-reference-able with the same results in each level.
- *Timeliness:* The data should be up to date.

2.5.2   Data Analysis Methods After Gathering the Required Data

The data gathered will be analyzed by the suitable software that is as follows:

- *Excel:* Used in data calculations related to the respondents. This software can help us to be assured that our gained data is accurate and has the required relevancy. Data that have too much variance or are very low or heterogeneous with other data can be easily identified in the visual display of this software. Moreover, for assessing the completeness of data, we can use the maximum likelihood estimation in this software.
- *SPSS:* Used in reliability calculations. This software can help us to be assured that our gained data is the desired consistency.
- *Mini Tab:* This software will be applied to the application of quality engineering techniques in the different sections of the circular supply chain management elements (This software will be used to design our model).

2.6   Creating an Adequate Literature Review or Research Background

The use of fuzzy techniques in various topics of science and engineering has been considered for more than 50 years. But the application of these techniques in various aspects of the supply chain has been very serious in the first two decades of the twenty-first century. In fact, the effectiveness and efficiency of fuzzy techniques have been proven to more researchers since the beginning of the twenty-first century.

This theory is a generalization of the classical theory of sets in mathematics, which was first presented by Professor Lotfizadeh. In the classical theory of sets, an element is either a member of the set or not. In fact, the membership of the elements follows a pattern of zero and one. But the theory of fuzzy sets expands this concept and proposes graded membership in such a way that an element can be a member of a set to some degree—and not completely. In this theory, the membership of the members of the set is determined by the function $U(x)$, where x represents a specific member and U is a fuzzy function that determines the degree of membership of x in the corresponding set, and its value is between zero and one, which is with $\mu(x)$. It is shown $A = \{(x, \mu(x)) \,|\, x \in U\}$. In other words, $U(x)$ is a mapping of x values to numerical values between zero and one. To facilitate and make fuzzy numbers practical, special fuzzy numbers are used in calculations. These special numbers are in the form of bells, triangles, trapezoids, etc. Among them, triangular fuzzy numbers (TFN) have been used more. A triangular fuzzy number can be represented as three ordered $x = (x_1, x_2, x_3)$ where x_1 and x_3 are the lower and upper limits and x_2 is the middle value, and $x_1 < x_2 < x_3$.

2.6.1   The Advantages of Using Fuzzy Techniques in Research Findings

- Fuzzy techniques are very useful from a conceptual point of view. The mathematical concepts of fuzzy techniques are very acceptable and reasonable.
- Fuzzy techniques are flexible. It is beneficial for researchers to add more functionality to any given system without starting over. It means that these fuzzy techniques are completely retrieval.
- Fuzzy techniques are resistant to imprecise data. If you look carefully around you, you will see that everything around us is imprecise. After all, most things are imprecise even under close inspection. Fuzzy techniques flow this understanding into the process rather than removing it from any process.
- Fuzzy techniques can model non-linear functions with any degree of complexity. You can create a fuzzy system to match any set of input and output data in each organization under study by researchers.
- Fuzzy techniques can be made based on the experience of experts. Unlike other techniques, fuzzy techniques allow you to rely on the experiences of people who already understand your system.
- Fuzzy techniques can be combined with conventional quality engineering techniques including statistical and non-statistical. Fuzzy techniques do not necessarily replace conventional control techniques. In many cases, fuzzy techniques use them or help to implement them.

2.6.2 The Previous Applications of Fuzzy Techniques in Supply Chains Management

Fuzzy analytic hierarchy process (AHP) and fuzzy technique for order of preference by similarity to ideal solution (TOPSIS) methods have been used in one research. In this study, a three-phase methodology has been applied for identifying, prioritizing, and ranking both barriers and solutions. The first phase studied the current situation in the electronics industry of Thailand and identified reverse logistics (RL) practice barriers and solutions to solve these barriers. The second phase used fuzzy AHP to get weight of criteria and sub-criteria of barriers and prioritized barriers. The third phase applied fuzzy TOPSIS to prioritize and rank the solutions of RL practice (Pornwasin Sirisawat and Tossapol Kiatcharoenpol 2018).

A fuzzy expert system has been deployed to select supply chain strategies: lean, agile, or leagile. In the first step, the primary design and system assumptions have been explained. In the second step, the fuzzification of input and output has been described. In the third step, the construction of fuzzy rules has been expressed. This is one of the best applications related to fuzzy concepts (Esfandiari et al. 2021).

A supply chain performance model has been introduced based on fuzzy logic to predict performance based on causal relationships between metrics of the supply council operations reference (SCOR) model. Indeed, this study was among the strong studies in the beginning of applying fuzzy concepts in measuring the performance of supply chain networks (Ganga and Carpinetti 2011).

The combination of the fuzzy analytical hierarchy process and the SCOR model has been deployed in supply chain analysis. This research was the beginning of combining fuzzy concepts with other techniques in the analysis of supply chain networks. The important aspect of this study was to consider model that covered the important elements including: (1) Reliability, (2) Cost, (3) Responsiveness, (4) Agility, and (5) Asset. Indeed, the structural shortcomings of this model have been also resolved using the AHP method very thoroughly (Akbar Abbaspour 2019).

A hierarchy multiple criteria decision-making (MCDM) model based on fuzzy-sets theory is proposed to deal with the supplier selection problems in the supply chain system. According to the concept of the TOPSIS, a closeness coefficient is defined to determine the ranking order of all suppliers by calculating the distances to the both fuzzy positive-ideal solution (FPIS) and fuzzy negative-ideal solution (FNIS) simultaneously (Chen et al. 2006).

2.6.3 The Previous Methods in the Performance Measurement of CSCM

The most current models in the performance measurement of circular supply chain management are as follows:

- Measuring the Product Supply Chain in CSCM (Potting et al. 2017).
- Measuring a Lifecycle Framework in CSCM (Maestrini et al. 2018).
- The Role Impact of Business Models in CSCM (Geissdoerfer et al. 2018).
- The Assessment of CSCM Potential at Territorial Level (Bassi et al. 2021).
- The Understanding of Barriers to CSCM Implementation (Ayati et al. 2022).
- The Multi-level and Scalable Performance Measurement in CSCM (Cagno et al. 2022).

The last definition of performance measurement in circular supply chain management is based on a balancing feedback loop at least in three papers (Vegter et al. 2020, 2021, 2023).

2.7   Highlighting the Applied Theories in the Previous Studies by Scholars

The applied theories that have been explained in the previous section have been shown in Table 2.1. The highlighting in this subsection indicates that there is no comprehensive approach to cover the combination of fuzzy techniques and quality

Table 2.1 Highlighting the applied theories in the previous studies

Author(s)	Application in CSCM	Applied theory	Year
Farooque et al.	The Definition of Circular SCM	Analytical	2019
Rostamkhani and Ramayah	Circular Supply Chain Management	Analytical	2022
Sirisawat and Kiatcharoenpol	The Fuzzy AHP + The Fuzzy TOPSIS	Analytical	2018
Esfandiari et al.	Fuzzy Expert System in Lean and Agile	Analytical	2021
Ganga and Carpinetti	Supply Council Operations Reference	Analytical	2011
Akbar Abbaspour	The Fuzzy AHP + The SCOR model	Analytical	2019
Chen et al.	MCDM model + Fuzzy-sets theory	Analytical	2006
Potting et al.	Measuring the Product Supply Chain	Analytical	2017
Maestrini et al.	Measuring a Lifecycle Framework	Analytical	2018
Geissdoerfer et al.	The Business Models in Circular SCM	Analytical	2018
Bassi et al.	The CSCM Potential at Territorial Level	Analytical	2021
Ayati et al.	The Barriers to CSCM Implementation	Analytical	2022
Cagno et al.	The Multi-level and Scalable Performance	Analytical	2022
Vegter et al.	The definition of performance measurement in circular supply chain management based on a balancing feedback loop	Analytical + Numerical	2020
Vegter et al.			2021
Vegter et al.			2023

engineering techniques for reaching circular thinking in supply chain management before the proposed model in the presented book. The previous models have tried to cover some aspects of this issue and could not achieve a competitive position in this research.

2.8   Conclusion

In this chapter, we paid attention to the current gap in the previous studies because they have incomplete information about the relationship between fuzzy techniques and the performance measurement of circular supply chain management elements. Moreover, there is no report to show the application of statistical methods as a facilitator tool for fuzzy techniques. Also, productivity and sustainability concepts have not been considered as a final goal in all previous studies. Therefore, at the end of this chapter, we have the required framework of the research structure and we know exactly what to look for in the coming chapters. So, we direct the attention of the readers of the book to four key elements as follows:

1. Measuring the performance of the circular supply chain management system is the final goal of the research in this book.
2. The statistical techniques such as the design of experiment (DOE) and the statistical process control (SPC) as the facilitating tools.
3. The selected fuzzy techniques (FAHP or FANP or FDEMATEL or Mixed Mode of Fuzzy Techniques) for measuring circular supply chain management performance with the facilitating role of statistical techniques (DOE +SPC).
4. Variance is a fairly common measure of uncertainty. Entropy is also often used; in some cases, they are equivalent. This book does not go into some detail on the topic, but we considered it when we received the experts' scores in the matrices related to all elements in circular thinking in single-to-single supply chain management components.

References

Abbaspour, A. (2019). Supply chain analysis and improvement by using the SCOR model and Fuzzy AHP: a case study. *International Journal of Industrial Engineering & Management Science, 6*(2), 51–73.

Ayati, M. S., Shekarian, E., Majava, J., & Wæhrens, B. V. (2022). Toward a circular supply chain: understanding barriers from the perspective of recovery approaches. *Journal of Cleaner Production, 131775.*

Bassi, A. M., Bianchi, M., Guzzetti, M., Pallaske, G., & Tapia, C. (2021). Improving the understanding of circular economy potential at territorial level using systems thinking. *Sustainable Production and Consumption, 27,* 128–140.

Cagno, E., Negri, M., Neri, A., & Giambone, M. (2022). One framework to rule them all: an integrated, multi-level and scalable performance measurement framework of sustainability, circular economy and industrial symbiosis. *Sustainable Production and Consumption, 35,* 55–71.

Chen, C. T., Lin, C. T., & Huang, S. F. (2006). A fuzzy approach for supplier evaluation and selection in supply chain management. *International Journal of Production Economics, 102*(2), 289–301.

Esfandiari, N., Moradi, M., & Golmohammadi, A. M. (2021). A fuzzy expert system to select a supply chain strategy: lean, agile or leagile. *Journal of Quality Engineering and Production Optimization, 6*(2), 201–218.

Ganga, G. M. D., & Carpinetti, L. C. R. (2011). A fuzzy logic approach to supply chain performance management. *International Journal of Production Economics, 134,* 177–187.

Geissdoerfer, M., Morioka, S. N., Carvalho, M. M., & Evans, S. (2018). Business models and supply chains for the circular economy. *Journal of Cleaner Production, 190,* 712–721.

Maestrini, V., Luzzini, D., Caniato, F., Maccarrone, P., & Ronchi, S. (2018). Measuring supply chain performance: a lifecycle framework and a case study. *International Journal of Operations & Production Management, 38*(4), 934–956.

Potting, J., Hekkert, M. P., Worrell, E., & Hanemaaijer, A. (2017). *Circular economy: measuring innovation in the product chain* (p. 2544). PBL Netherlands Assessment Agency.

Sirisawat, P., & Kiatcharoenpol, T. (2018). Fuzzy AHP-TOPSIS approaches to prioritizing solutions for reverse logistics barriers. *Computers & Industrial Engineering, 117,* 303–318.

Vegter, D., Hillegersberg, J. V., & Olthaar, M. (2020). Supply chains in circular business models: processes and performance objectives. *Resources, Conservation and Recycling, 162,* 105046.

Vegter, D., Hillegersberg, J. V., & Olthaar, M. (2021). Performance measurement systems for circular supply chain management: current state of development. *Sustainability, 13*(21), 12082.

Vegter, D., Hillegersberg, J. V., & Olthaar, M. (2023). Performance measurement system for circular supply chain management. *Sustainable Production and Consumption, 36,* 171–183.

Chapter 3
Numerical Application in the Sample Organization

3.1 Introduction

In the third chapter, the main core of proposed model in the book has been implemented. This chapter consists of three basic pillars. In the first step, we seek the best fuzzy techniques from the popular fuzzy techniques utilizing the design of experiments. In the second step, we conceptualize the circular thinking in the supply chain management elements and use the selected fuzzy technique in the performance assessment of all subsections. In the third step, we utilize the selected fuzzy technique by ZM-Charts to assess the relevant data to be sure that they are under control. In the final step, we introduce all required indices and sub-indices related to the management of productivity and sustainability processes in the sample organization. The important issue in the analysis of this chapter is to have a circular approach in each element of the supply chain. The strength of this book is in the three sections described in this chapter, and we must note that the output of each section is the input of the next section. Figure 3.1 shows a summary of the contents of this chapter.

3.2 Selecting the Best Fuzzy Technique by the Design of Experiment

The compatibility index is defined as follows:

$$C_{\mathrm{I}} = \frac{\lambda_{\max} - n}{n - 1} \tag{3.1}$$

where n is the number of fuzzy techniques available in the circular supply chain management elements and $\lambda_{\max}$ is the compatibility average of indices vector. Choosing the best fuzzy technique among the introduced fuzzy techniques is the

© The Author(s), under exclusive license to Springer Nature Singapore Pte Ltd. 2024
R. Rostamkhani, T. Ramayah, *Navigating Circular Supply Chains*, SpringerBriefs in Operations Management, https://doi.org/10.1007/978-981-97-4704-7_3

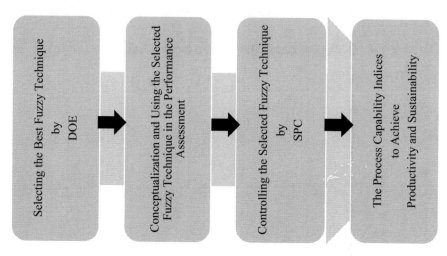

Fig. 3.1 A summary of this chapter

Table 3.1 The compatibility average of indices in circular SCM for four fuzzy methods

Fuzzy methods	λ_{max}				
The indices in circular SCM	FAHP	FDEMATEL	FANP	Mixed mode (FDEMATEL + FANP)	Y_{io}
I_1	56.52	64.15	73.22	85.75	283.64
I_2	54.22	63.12	75.62	82.33	275.29
I_3	56.42	65.14	76.75	87.25	285.56
I_4	48.38	67.28	76.95	88.12	280.73
I_5	52.38	65.73	77.65	90.33	286.09
I_6	62.26	66.52	75.25	77.28	281.31
I_7	55.52	55.15	66.62	85.26	262.55
I_8	45.12	53.12	65.49	78.37	242.10
I_9	49.83	55.28	68.98	72.28	246.37
Y_{oj}	480.65	555.49	656.53	750.97	2443.64

output of the design of experiments in our study. In fact, we want to determine whether the difference between fuzzy techniques in the analysis of circular supply chain data makes a significant difference, and if the answer is positive, which fuzzy technique has the best situation in analyzing these data. Table 3.1 shows the total framework in this step. (It should be noted that all indices are in terms of percentage.)

The relevant calculations are as follows:

$$SS_{Indices} = \frac{1}{4}(665,815.97) - \frac{(2443.64)^2}{36} = 582.43 \rightarrow MS_{indices} = \frac{582.43}{9-1} = 72.80$$

$$SS_{fuzzy\ techniques} = \frac{1}{9}(1,534,581.14) - \frac{(2443.64)^2}{36} = 4637.45 \rightarrow$$

$$MS_{fuzzy\ techniques} = \frac{4637.45}{4-1} = 1545.82$$

$$\text{SS}_{\text{error}} = 171,460.23 - \frac{5,971,376.45}{36} = 5588.66 \rightarrow \text{MS}_{\text{error}} = \frac{5588.66}{(9-1)(4-1)}$$

$$= 232.86$$

Therefore:

$F_{\text{indices}} = \frac{72.80}{232.86} = 0.31$	$F_{\text{fuzzy techniques}} = \frac{1545.82}{232.86} = 6.64$

The Analysis of Circular Supply Chain Management Data:
There are four categories in the analysis of each scientific data that are as follows:

1. Predictive Analysis: What could happen
2. Descriptive Analysis: What has already happened
3. Prescriptive Analysis: What should happen in the future
4. Multiple Analysis: What has already happened + what could happen + what should happen in the future

The relevant analysis in this chapter is the multiple states. Previous experiences have shown that this compatibility index vector provides more tangible results with fuzzy combined techniques. In addition, this analysis is what we expect and want to happen in the future.

$$F_{\text{indices}} = 0.31 \qquad F_{0.1,8,24} = 1.94 \qquad \rightarrow 0.31 < 1.94$$

The *indices* do not create any significant difference. This means that the difference in the circular chain management indices does not make a significant difference in the assessment.

$$F_{\text{fuzzy techniques}} = 6.64 \qquad F_{0.1,3,24} = 2.33 \qquad \rightarrow 6.64 > 2.33$$

The type of *fuzzy technique* creates any significant difference. This means that the difference in the fuzzy technique makes a significant difference in the assessment.

Figure 3.2 shows the relevant values related to λ_{max} for each fuzzy technique by the circular supply chain management elements.

The maximum compatibility average of indices vector belongs to the mixed mode (FDEMATEL + FANP). It means that the combination of FDEMATEL and FANP can create the maximum compatibility for the circular supply chain management data (Table 3.2).

The characteristics of the two fuzzy techniques are combined in this mode and provide a fuzzy technique that covers the weaknesses of both fuzzy techniques.

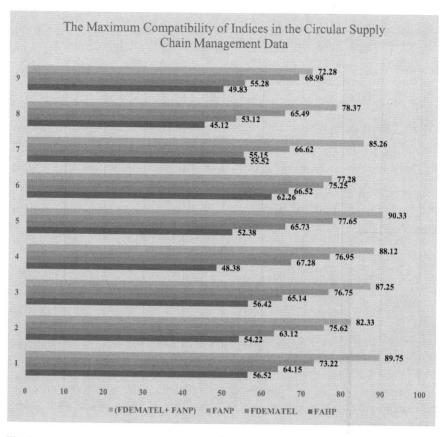

Fig. 3.2 The values of $\lambda_{\max}$ for each fuzzy technique by the CSCM elements

Table 3.2 $\lambda_{\max}$ in the circular supply chain management data for four fuzzy methods

FAHP	FDEMATEL	FANP	FDEMATEL + FANP
$\lambda_{\max} = 53.405$	$\lambda_{\max} = 61.72$	$\lambda_{\max} = 72.95$	$\lambda_{\max} = 83.44$

3.3 The Conceptualization and Using the (FDEMATEL + FANP) in the PA[1]

After choosing the best fuzzy technique with the design of experiments, we proceed to conceptualize circular thinking in the supply chain management elements, and in the next step, we use the selected fuzzy technique (FDEMATEL+FANP) in analyzing the performance of the circular supply chain structure.

---

[1] Performance Assessment.

Conceptualization of the Circular Thinking in SCM

Application of FDEMATEL in the Performance Assessment of CSCM(3 Steps)

Application of FANP in the Performance Assessment of CSCM(Final Result)

Fig. 3.3 The main core of proposed conceptualization and application of the selected fuzzy technique

In general, conceptualization is the process of specifying what we mean when we use particular terms. The conceptual frameworks can be written or visual and are generally developed based on a literature review of existing studies about each topic.

In this subsection, we want to explain the related circular thinking in all supply chain management elements and want to describe how we can use the mixed fuzzy techniques (FDEMATEL+FANP) for their performance assessment.

Figure 3.3 shows the order that we want to apply in this subsection.

The supply chain management elements' performance is assessed in this subsection after conceptualizing the circular thinking in them (Appendix has shown all required calculations in each element in EXCEL).

3.3.1   The Conceptualization of Retrieving of Customers' Need

Figure 3.4 shows the full demonstration and the conceptualization of retrieving of customers' need.

3.3.2   The Mixed Fuzzy Method (FDEMATEL + FANP) in the PA of Customers' Need

Tables 3.3, 3.4, 3.5, and 3.6 show the application of mixed fuzzy method (FDEMATEL + FANP) in the performance assessment of customers' need in the CSCM elements.

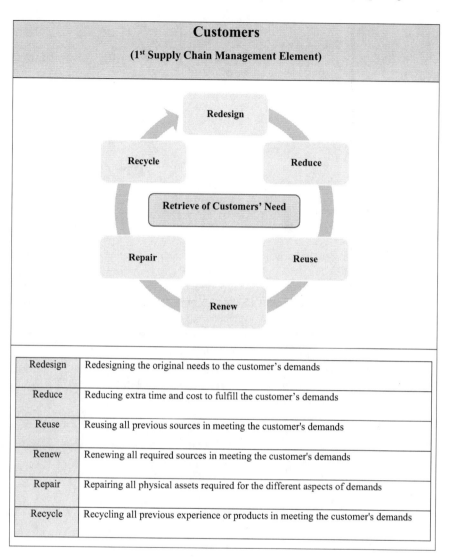

Fig. 3.4 The conceptualization of retrieving of customers' need

Table 3.3 The Fuzzy DEMATEL Matrix for making circular thinking in the customer field (1)

	Redesign	Reduce	Reuse	Renew	Repair	Recycle
Redesign	–	1	3	4	0	2
Reduce	0	–	3	4	1	1
Reuse	2	2	–	3	0	2
Renew	3	3	2	–	1	1
Repair	1	2	1	3	–	2
Recycle	0	2	2	2	0	–

Table 3.4 The Fuzzy DEMATEL Matrix for making circular thinking in the customer field (2)

	Redesign	Reduce	Reuse	Renew	Repair	Recycle
Redesign	–	0.20	0.60	0.80	0.00	0.40
Reduce	0.00	–	0.60	0.80	0.20	0.20
Reuse	0.40	0.40	–	0.60	0.00	0.40
Renew	0.60	0.60	0.40	–	0.20	0.20
Repair	0.20	0.40	0.20	0.60	–	0.40
Recycle	0.00	0.40	0.40	0.40	0.00	–

Table 3.5 The priority percentage in the Fuzzy DEMATEL Matrix in the customer field

	Redesign	Reduce	Reuse	Renew	Repair	Recycle
Redesign	–	10.0%	27.3%	25.0%	0.0%	25.0%
Reduce	0.0%	–	27.3%	25.0%	50.0%	12.5%
Reuse	33.3%	20.0%	–	18.8%	0.0%	25.0%
Renew	50.0%	30.0%	18.2%	–	50.0%	12.5%
Repair	16.7%	20.0%	9.1%	18.8%	–	25.0%
Recycle	0.0%	20.0%	18.2%	12.5%	0.0%	–

Table 3.6 The calculation of metrics using the FANP Method in the customer field

N	Name	Degree of importance (The total priority percentage)	Degree of influence (TPP-100)
1	Renew	160.7	60.7
2	Reduce	114.8	14.8
3	Reuse	97.1	-2.9

Therefore, the three important factors in evaluating the performance of the studied circular supply chain in the customer field are as follows in order of priority:
1—Renew, 2—Reduce, 3—Reuse

3.3.3   The Conceptualization of Retrieving of Demand Prediction

Figure 3.5 shows the full demonstration and the conceptualization of retrieving of demand prediction.

3.3.4   The Mixed Fuzzy Method (FDEMATEL + FANP) in the PA of Demand Prediction

Tables 3.7, 3.8, 3.9, and 3.10 show the application of mixed fuzzy method (FDEMATEL + FANP) in the performance assessment of demand predicting in the CSCM elements.

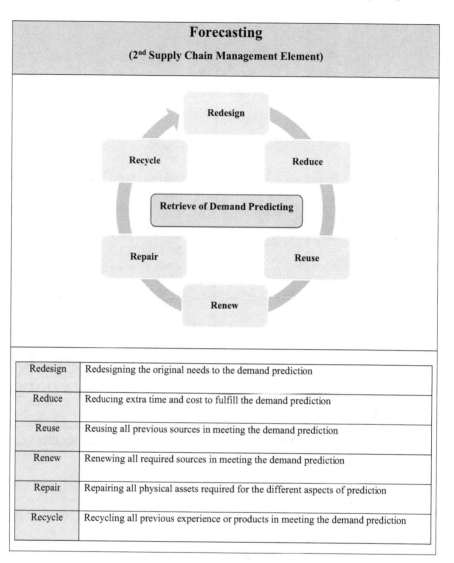

Fig. 3.5 The conceptualization of retrieving of demand prediction

Table 3.7 The Fuzzy DEMATEL Matrix for making circular thinking in the forecasting field (1)

	Redesign	Reduce	Reuse	Renew	Repair	Recycle
Redesign	–	1	4	4	1	4
Reduce	0	–	4	4	0	3
Reuse	2	2	–	3	1	4
Renew	3	3	2	–	0	3
Repair	1	2	1	3	–	3
Recycle	0	2	2	2	1	–

Table 3.8 The Fuzzy DEMATEL Matrix for making circular thinking in the forecasting field (2)

	Redesign	Reduce	Reuse	Renew	Repair	Recycle
Redesign	–	0.20	0.80	0.80	0.20	0.80
Reduce	0.00	–	0.80	0.80	0.00	0.60
Reuse	0.40	0.40	–	0.60	0.20	0.80
Renew	0.60	0.60	0.40	–	0.00	0.60
Repair	0.20	0.40	0.20	0.60	–	0.60
Recycle	0.00	0.40	0.40	0.40	0.20	–

Table 3.9 The priority percentage in the Fuzzy DEMATEL Matrix in the forecasting field

	Redesign	Reduce	Reuse	Renew	Repair	Recycle
Redesign	–	10.0%	30.8%	25.0%	33.3%	23.5%
Reduce	0.0%	–	30.8%	25.0%	0.0%	17.6%
Reuse	33.3%	20.0%	–	18.8%	33.3%	23.5%
Renew	50.0%	30.0%	15.4%	–	0.0%	17.6%
Repair	16.7%	20.0%	7.7%	18.8%	–	17.6%
Recycle	0.0%	20.0%	15.4%	12.5%	33.3%	–

Table 3.10 The calculation of metrics using the FANP Method in the forecasting field

N	Name	Degree of importance (The total priority percentage)	Degree of influence (TPP-100)
1	Reuse	128.9	28.9
2	Redesign	122.6	22.6
3	Renew	113.0	13.0

Therefore, the three important factors in evaluating the performance of the studied circular supply chain in the forecasting field are as follows in order of priority:
1—Reuse, 2—Redesign, 3—Renew

3.3.5 The Conceptualization of Retrieving of Required Specifications

Figure 3.6 shows the full demonstration and the conceptualization of retrieving of required specification.

3.3.6 The Mixed Fuzzy Method (FDEMATEL + FANP) in the PA of Required Specifications

Tables 3.11, 3.12, 3.13, and 3.14 show the application of mixed fuzzy method (FDEMATEL + FANP) in the performance assessment of required specifications in the CSCM elements.

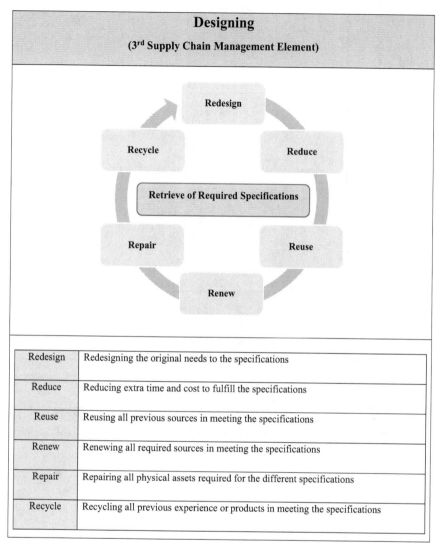

Fig. 3.6 The conceptualization of retrieving of required specification

Table 3.11 The Fuzzy DEMATEL Matrix for making circular thinking in the designing field (1)

	Redesign	Reduce	Reuse	Renew	Repair	Recycle
Redesign	–	4	4	4	0	3
Reduce	4	–	3	2	1	1
Reuse	4	2	–	3	0	2
Renew	3	3	2	–	1	1
Repair	3	0	1	0	–	1
Recycle	0	2	2	2	0	–

Table 3.12 The Fuzzy DEMATEL Matrix for making circular thinking in the designing field (2)

	Redesign	Reduce	Reuse	Renew	Repair	Recycle
Redesign	–	0.80	0.80	0.80	0.00	0.60
Reduce	0.80	–	0.60	0.40	0.20	0.20
Reuse	0.80	0.40	–	0.60	0.00	0.40
Renew	0.60	0.60	0.40	–	0.20	0.20
Repair	0.60	0.00	0.20	0.00	–	0.20
Recycle	0.00	0.40	0.40	0.40	0.00	–

Table 3.13 The priority percentage in the Fuzzy DEMATEL Matrix in the designing field

	Redesign	Reduce	Reuse	Renew	Repair	Recycle
Redesign	–	36.4%	33.3%	36.4%	0.0%	37.5%
Reduce	28.6%	–	25.0%	18.2%	50.0%	12.5%
Reuse	28.6%	18.2%	–	27.3%	0.0%	25.0%
Renew	21.4%	27.3%	16.7%	–	50.0%	12.5%
Repair	21.4%	0.0%	8.3%	0.0%	–	12.5%
Recycle	0.0%	18.2%	16.7%	18.2%	0.0%	–

Table 3.14 The calculation of metrics using the FANP Method in the designing field

N	Name	Degree of importance (The total priority percentage)	Degree of influence (TPP-100)
1	Redesign	143.6	43.6
2	Reduce	134.3	34.3
3	Renew	127.9	27.9

Therefore, the three important factors in evaluating the performance of the studied circular supply chain in the designing field are as follows in order of priority:
1—Redesign, 2—Reduce, 3—Renew

3.3.7   The Conceptualization of Retrieving of Quality Control

Figure 3.7 shows the full demonstration and the conceptualization of retrieving of quality control.

3.3.8   The Mixed Fuzzy Method (FDEMATEL + FANP) in the PA of Quality Control

Tables 3.15, 3.16, 3.17, and 3.18 show the application of mixed fuzzy method (FDEMATEL + FANP) in the performance assessment of quality control in the CSCM elements.

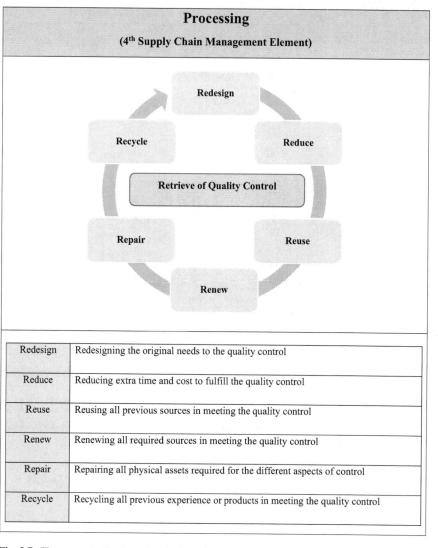

Fig. 3.7 The conceptualization of retrieving of quality control

Table 3.15 The Fuzzy DEMATEL Matrix for making circular thinking in the processing field (1)

	Redesign	Reduce	Reuse	Renew	Repair	Recycle
Redesign	–	1	1	1	1	1
Reduce	1	–	3	3	4	4
Reuse	1	2	–	3	4	2
Renew	3	3	2	–	3	3
Repair	2	2	2	3	–	2
Recycle	3	2	3	4	2	–

Table 3.16 The Fuzzy DEMATEL Matrix for making circular thinking in the processing field (2)

	Redesign	Reduce	Reuse	Renew	Repair	Recycle
Redesign	–	0.20	0.20	0.20	0.20	0.20
Reduce	0.20	–	0.60	0.60	0.80	0.80
Reuse	0.20	0.40	–	0.60	0.80	0.40
Renew	0.60	0.60	0.40	–	0.60	0.60
Repair	0.40	0.40	0.40	0.60	–	0.40
Recycle	0.60	0.40	0.60	0.80	0.40	–

Table 3.17 The priority percentage in the Fuzzy DEMATEL Matrix in the processing field

	Redesign	Reduce	Reuse	Renew	Repair	Recycle
Redesign	–	10.0%	9.1%	7.1%	7.1%	8.3%
Reduce	10.0%	–	27.3%	21.4%	28.6%	33.3%
Reuse	10.0%	20.0%	–	21.4%	28.6%	16.7%
Renew	30.0%	30.0%	18.2%	–	21.4%	25.0%
Repair	20.0%	20.0%	18.2%	21.4%	–	16.7%
Recycle	30.0%	20.0%	27.3%	28.6%	14.3%	–

Table 3.18 The calculation of metrics using the FANP Method in the processing field

N	Name	Degree of importance (The total priority percentage)	Degree of influence (TPP-100)
1	Renew	124.6	24.6
2	Reduce	120.6	20.6
3	Recycle	120.1	20.1

Therefore, the three important factors in evaluating the performance of the studied circular supply chain in the processing field are as follows in order of priority:
1—Renew, 2—Reduce, 3—Recycle

3.3.9   The Conceptualization of Retrieving of Inventory Management

Figure 3.8 shows the full demonstration and the conceptualization of retrieving of inventory management.

3.3.10   The Mixed Fuzzy Method (FDEMATEL + FANP) in the PA of Inventory Management

Tables 3.19, 3.20, 3.21, and 3.22 show the application of mixed fuzzy method (FDEMATEL + FANP) in the performance assessment of inventory management in the CSCM elements.

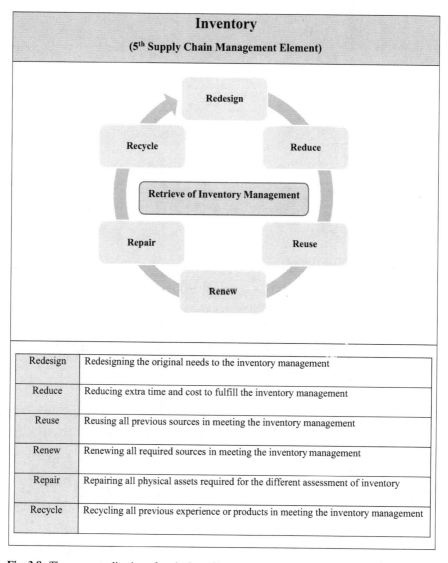

Fig. 3.8 The conceptualization of retrieving of inventory management

3.3.11 The Conceptualization of Retrieving of Suppliers Evaluation

Figure 3.9 shows the full demonstration and the conceptualization of retrieving of suppliers evaluation.

Table 3.19 The Fuzzy DEMATEL Matrix for making circular thinking in the inventory field (1)

	Redesign	Reduce	Reuse	Renew	Repair	Recycle
Redesign	–	1	0	1	0	1
Reduce	0	–	1	2	2	1
Reuse	2	2	–	4	4	4
Renew	4	4	4	–	3	3
Repair	2	2	1	1	–	1
Recycle	4	4	4	3	3	–

Table 3.20 The Fuzzy DEMATEL Matrix for making circular thinking in the inventory field (2)

	Redesign	Reduce	Reuse	Renew	Repair	Recycle
Redesign	–	0.20	0.00	0.20	0.00	0.20
Reduce	0.00	–	0.20	0.40	0.40	0.20
Reuse	0.40	0.40	–	0.80	0.80	0.80
Renew	0.80	0.80	0.80	–	0.60	0.60
Repair	0.40	0.40	0.20	0.20	–	0.20
Recycle	0.80	0.80	0.80	0.60	0.60	–

Table 3.21 The priority percentage in the Fuzzy DEMATEL Matrix in the inventory field

	Redesign	Reduce	Reuse	Renew	Repair	Recycle
Redesign	–	7.7%	0.0%	9.1%	0.0%	10.0%
Reduce	0.0%	–	10.0%	18.2%	16.7%	10.0%
Reuse	16.7%	15.4%	–	36.4%	33.3%	40.0%
Renew	33.3%	30.8%	40.0%	–	25.0%	30.0%
Repair	16.7%	15.4%	10.0%	9.1%	–	10.0%
Recycle	33.3%	30.8%	40.0%	27.3%	25.0%	–

Table 3.22 The calculation of metrics using the FANP Method in the inventory field

N	Name	Degree of importance (The total priority percentage)	Degree of influence (TPP-100)
1	Renew	159.1	59.1
2	Recycle	156.4	56.4
3	Reuse	141.7	41.7

Therefore, the three important factors in evaluating the performance of the studied circular supply chain in the inventory field are as follows in order of priority:
1—Renew, 2—Recycle, 3—Reuse

3.3.12   The Mixed Fuzzy Method (FDEMATEL + FANP) in the PA of Suppliers Evaluation

Tables 3.23, 3.24, 3.25, and 3.26 show the application of mixed fuzzy method (FDEMATEL + FANP) in the performance assessment of suppliers evaluation in the CSCM elements.

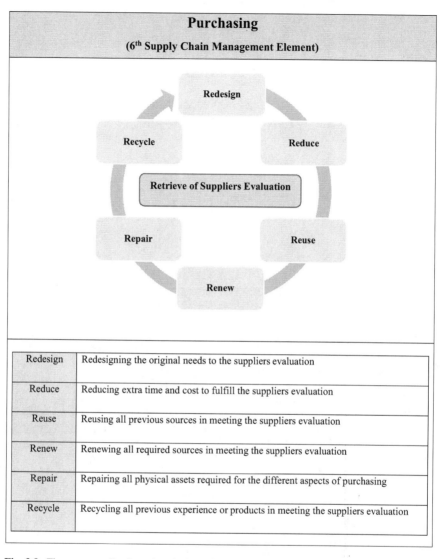

Purchasing
(6th Supply Chain Management Element)

Redesign	Redesigning the original needs to the suppliers evaluation
Reduce	Reducing extra time and cost to fulfill the suppliers evaluation
Reuse	Reusing all previous sources in meeting the suppliers evaluation
Renew	Renewing all required sources in meeting the suppliers evaluation
Repair	Repairing all physical assets required for the different aspects of purchasing
Recycle	Recycling all previous experience or products in meeting the suppliers evaluation

Fig. 3.9 The conceptualization of retrieving of suppliers evaluation

Table 3.23 The Fuzzy DEMATEL Matrix for making circular thinking in the purchasing field (1)

	Redesign	Reduce	Reuse	Renew	Repair	Recycle
Redesign	–	1	0	2	1	0
Reduce	0	–	3	3	3	3
Reuse	2	2	–	4	4	4
Renew	2	2	2	–	2	2
Repair	1	1	1	1	–	0
Recycle	3	3	2	2	3	–

Table 3.24 The Fuzzy DEMATEL Matrix for making circular thinking in the purchasing field (2)

	Redesign	Reduce	Reuse	Renew	Repair	Recycle
Redesign	–	0.20	0.00	0.40	0.20	0.00
Reduce	0.00	–	0.60	0.60	0.60	0.60
Reuse	0.40	0.40	–	0.80	0.80	0.80
Renew	0.40	0.40	0.40	–	0.40	0.40
Repair	0.20	0.20	0.20	0.20	–	0.00
Recycle	0.60	0.60	0.40	0.40	0.60	–

Table 3.25 The priority percentage in the Fuzzy DEMATEL Matrix in the purchasing field

	Redesign	Reduce	Reuse	Renew	Repair	Recycle
Redesign	–	11.1%	0.0%	16.7%	7.7%	0.0%
Reduce	0.0%	–	37.5%	25.0%	23.1%	33.3%
Reuse	25.0%	22.2%	–	33.3%	30.8%	44.4%
Renew	25.0%	22.2%	25.0%	–	15.4%	22.2%
Repair	12.5%	11.1%	12.5%	8.3%	–	0.0%
Recycle	37.5%	33.3%	25.0%	16.7%	23.1%	–

Table 3.26 The calculation of metrics using the FANP Method in the purchasing field

N	Name	Degree of importance (The total priority percentage)	Degree of influence (TPP-100)
1	Reuse	155.8	55.8
2	Recycle	135.6	35.6
3	Reduce	118.9	18.9

Therefore, the three important factors in evaluating the performance of the studied circular supply chain in the purchasing field are as follows in order of priority:
1—Reuse, 2—Recycle, 3—Reduce

3.3.13   The Conceptualization of Retrieving of Suppliers' Quality

Figure 3.10 shows the full demonstration and the conceptualization of retrieving of suppliers' quality.

3.3.14   The Mixed Fuzzy Method (FDEMATEL + FANP) in the PA of Suppliers' Quality

Tables 3.27, 3.28, 3.29, and 3.30 show the application of mixed fuzzy method (FDEMATEL + FANP) in the performance assessment of suppliers' quality in the CSCM elements.

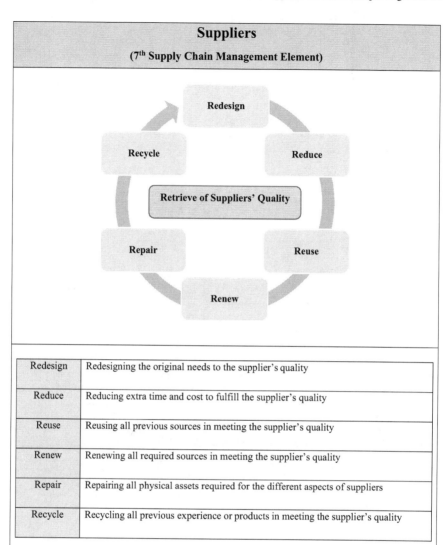

Fig. 3.10 The conceptualization of retrieving of suppliers' quality

Table 3.27 The Fuzzy DEMATEL Matrix for making circular thinking in the suppliers field (1)

	Redesign	Reduce	Reuse	Renew	Repair	Recycle
Redesign	–	1	0	2	0	1
Reduce	0	–	3	2	3	2
Reuse	1	1	–	1	1	1
Renew	3	3	2	–	1	1
Repair	1	2	1	3	–	2
Recycle	0	2	2	2	0	–

Table 3.28 The Fuzzy DEMATEL Matrix for making circular thinking in the suppliers field (2)

	Redesign	Reduce	Reuse	Renew	Repair	Recycle
Redesign	–	0.20	0.00	0.40	0.00	0.20
Reduce	0.00	–	0.60	0.40	0.60	0.40
Reuse	0.20	0.20	–	0.20	0.20	0.20
Renew	0.60	0.60	0.40	–	0.20	0.20
Repair	0.20	0.40	0.20	0.60	–	0.40
Recycle	0.00	0.40	0:40	0.40	0.00	–

Table 3.29 The priority percentage in the Fuzzy DEMATEL Matrix in the suppliers field

	Redesign	Reduce	Reuse	Renew	Repair	Recycle
Redesign	–	11.1%	0.0%	20.0%	0.0%	14.3%
Reduce	0.0%	–	37.5%	20.0%	60.0%	28.6%
Reuse	20.0%	11.1%	–	10.0%	20.0%	14.3%
Renew	60.0%	33.3%	25.0%	–	20.0%	14.3%
Repair	20.0%	22.2%	12.5%	30.0%	–	28.6%
Recycle	0.0%	22.2%	25.0%	20.0%	0.0%	–

Table 3.30 The calculation of metrics using the FANP Method in the suppliers field

N	Name	Degree of importance (The total priority percentage)	Degree of influence (TPP-100)
1	Renew	152.6	52.6
2	Reduce	146.1	46.1
3	Repair	113.3	13.3

Therefore, the three important factors in evaluating the performance of the studied circular supply chain in the suppliers field are as follows in order of priority:
1—Renew, 2—Reduce, 3—Repair

3.3.15 The Conceptualization of Retrieving of Location Determination

Figure 3.11 shows the full demonstration and the conceptualization of retrieving of location determination.

3.3.16 The Mixed Fuzzy Method (FDEMATEL + FANP) in the PA of Location Determination

Tables 3.31, 3.32, 3.33, and 3.34 show the application of mixed fuzzy method (FDEMATEL + FANP) in the performance assessment of location determination in the CSCM elements.

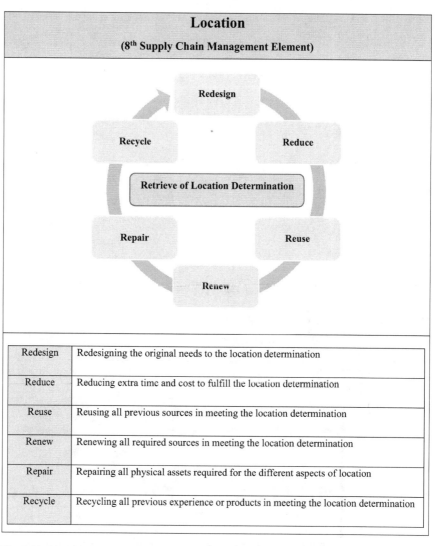

Fig. 3.11 The conceptualization of retrieving of location determination

Table 3.31 The Fuzzy DEMATEL Matrix for making circular thinking in the location field (1)

	Redesign	Reduce	Reuse	Renew	Repair	Recycle
Redesign	–	2	2	2	2	2
Reduce	0	–	3	3	3	3
Reuse	4	3	–	4	3	3
Renew	2	3	2	–	3	2
Repair	4	4	4	3	–	3
Recycle	3	2	1	3	2	–

Table 3.32 The Fuzzy DEMATEL Matrix for making circular thinking in the location field (2)

	Redesign	Reduce	Reuse	Renew	Repair	Recycle
Redesign	–	0.40	0.40	0.40	0.40	0.40
Reduce	0.00	–	0.60	0.60	0.60	0.60
Reuse	0.80	0.60	–	0.80	0.60	0.60
Renew	0.40	0.60	0.40	–	0.60	0.40
Repair	0.80	0.80	0.80	0.60	–	0.60
Recycle	0.60	0.40	0.20	0.60	0.40	–

Table 3.33 The priority percentage in the Fuzzy DEMATEL Matrix in the location field

	Redesign	Reduce	Reuse	Renew	Repair	Recycle
Redesign	–	14.3%	16.7%	13.3%	15.4%	15.4%
Reduce	0.0%	–	25.0%	20.0%	23.1%	23.1%
Reuse	30.8%	21.4%	–	26.7%	23.1%	23.1%
Renew	15.4%	21.4%	16.7%	–	23.1%	15.4%
Repair	30.8%	28.6%	33.3%	20.0%	–	23.1%
Recycle	23.1%	14.3%	8.3%	20.0%	15.4%	–

Table 3.34 The calculation of metrics using the FANP Method in the location field

N	Name	Degree of importance (The total priority percentage)	Degree of influence (TPP-100)
1	Repair	135.8	35.8
2	Reuse	125	25
3	Renew	91.9	−8.1

Therefore, the three important factors in evaluating the performance of the studied circular supply chain in the location field are as follows in order of priority:
1—Repair, 2—Reuse, 3—Renew

3.3.17   The Conceptualization of Retrieving of Movement and Storage

Figure 3.12 shows the full demonstration and the conceptualization of retrieving of movement and storage.

3.3.18   The Mixed Fuzzy Method (FDEMATEL + FANP) in the PA of Movement and Storage

Tables 3.35, 3.36, 3.37, and 3.38 show the application of mixed fuzzy method (FDEMATEL + FANP) in the performance assessment of movement and storage in the CSCM elements.

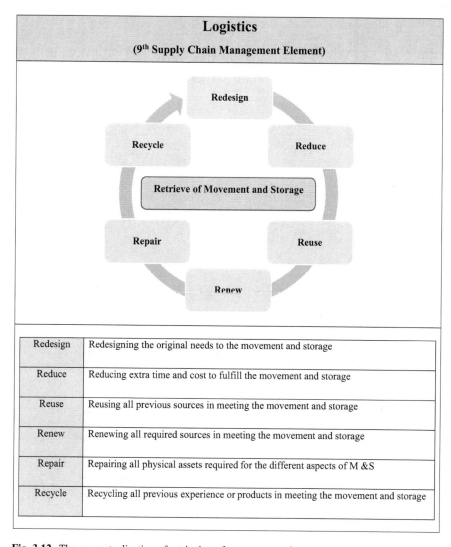

Fig. 3.12 The conceptualization of retrieving of movement and storage

Table 3.35 The Fuzzy DEMATEL Matrix for making circular thinking in the logistics field (1)

	Redesign	Reduce	Reuse	Renew	Repair	Recycle
Redesign	–	1	2	2	3	3
Reduce	0	–	2	2	2	2
Reuse	3	3	–	3	3	3
Renew	2	2	2	–	1	1
Repair	4	4	4	4	–	4
Recycle	0	1	1	2	2	–

Table 3.36 The Fuzzy DEMATEL Matrix for making circular thinking in the logistics field (2)

	Redesign	Reduce	Reuse	Renew	Repair	Recycle
Redesign	–	0.20	0.40	0.40	0.60	0.60
Reduce	0.00	–	0.40	0.40	0.40	0.40
Reuse	0.60	0.60	–	0.60	0.60	0.60
Renew	0.40	0.40	0.40	–	0.20	0.20
Repair	0.80	0.80	0.80	0.80	–	0.80
Recycle	0.00	0.20	0.20	0.40	0.40	–

Table 3.37 The Priority percentage in the Fuzzy DEMATEL Matrix in the logistics field

	Redesign	Reduce	Reuse	Renew	Repair	Recycle
Redesign	–	9.1%	18.2%	15.4%	27.3%	23.1%
Reduce	0.0%	–	18.2%	15.4%	18.2%	15.4%
Reuse	33.3%	27.3%	–	23.1%	27.3%	23.1%
Renew	22.2%	18.2%	18.2%	–	9.1%	7.7%
Repair	44.4%	36.4%	36.4%	30.8%	–	30.8%
Recycle	0.0%	9.1%	9.1%	15.4%	18.2%	–

Table 3.38 The calculation of metrics using the FANP Method in the logistics field

N	Name	Degree of importance (The total priority percentage)	Degree of influence (TPP-100)
1	Repair	178.7	78.7
2	Reuse	134	34
3	Redesign	93	−7

Therefore, the three important factors in evaluating the performance of the studied circular supply chain in the movement and storage field are as follows in order of priority:
1—Repair, 2—Reuse, 3—Redesign

3.4 Controlling the Extracted Data from Fuzzy Method by the Z-MR Charts

In this section, the Z-MR control charts show the checking of the selected fuzzy method (FDEMATEL+FANP) in this book in the different elements of circular supply chain management and whether they can create data under control or not.

3.4.1 The Application of the Z-MR Chart for Obtained Data (Customers' Need)

Table 3.39 and Fig. 3.13 show the extracted data in the domain of customers.

All data extracted from the selected fuzzy method in this domain regarding the considered target are under control.

Table 3.39 The extracted data from the selected fuzzy method in the customers' need

N	Name	Fuzzy method (FDEMATEL + FANP)	
		Target	Total priority
1	Redesign	1.00	0.87
2	Reduce	1.00	1.15
3	Reuse	1.00	0.97
4	Renew	1.00	1.61
5	Repair	1.00	0.90
6	Recycle	1.00	0.51

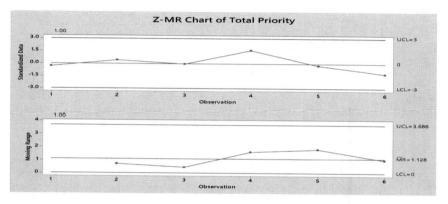

Fig. 3.13 The data demonstration in the domain of customers

Table 3.40 The extracted data from the selected fuzzy method in the demand prediction

N	Name	Fuzzy method (FDEMATEL + FANP)	
		Target	Total priority
1	Redesign	1.00	1.23
2	Reduce	1.00	0.73
3	Reuse	1.00	1.29
4	Renew	1.00	1.13
5	Repair	1.00	0.81
6	Recycle	1.00	0.81

3.4.2   The Application of the Z-MR Chart for Obtained Data (Demand Prediction)

Table 3.40 and Fig. 3.14 show the extracted data in the domain of forecasting.

All data extracted from the selected fuzzy method in this domain regarding the considered target are under control. Please pay attention to the sixth data that is located on one of the borders in moving range.

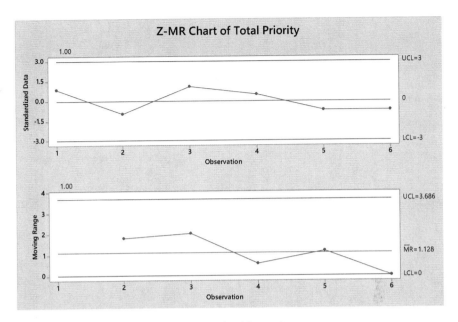

Fig. 3.14 The data demonstration in the domain of forecasting

			Fuzzy method (FDEMATEL + FANP)	
N		Name	Target	Total priority
1		Redesign	1.00	1.44
2		Reduce	1.00	1.34
3		Reuse	1.00	0.99
4		Renew	1.00	1.28
5		Repair	1.00	0.42
6		Recycle	1.00	0.53

Table 3.41 The extracted data from the selected fuzzy method in the required specifications

3.4.3 The Application of the Z-MR Chart for Obtained Data (Required Specification)

Table 3.41 and Fig. 3.15 show the extracted data in the domain of designing.

All data extracted from the selected fuzzy method in this domain regarding the considered target are under control.

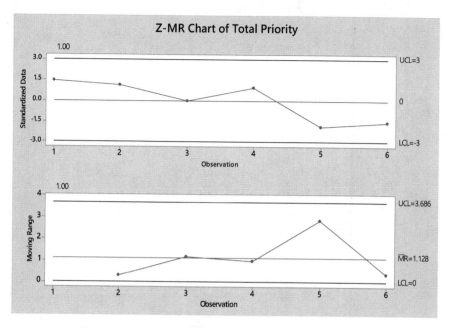

Fig. 3.15 The data demonstration in the domain of designing

Table 3.42 The extracted data from the selected fuzzy method in the quality control

N	Name	λ_{max} (FDEMATEL + FANP)	
		Target	Total priority
1	Redesign	1.00	0.42
2	Reduce	1.00	1.21
3	Reuse	1.00	0.97
4	Renew	1.00	1.25
5	Repair	1.00	0.96
6	Recycle	1.00	1.20

3.4.4   The Application of the Z-MR Chart for Obtained Data (Quality Control)

Table 3.42 and Fig. 3.16 show the extracted data in the domain of processing.

All data extracted from the selected fuzzy method in this domain regarding the considered target are under control.

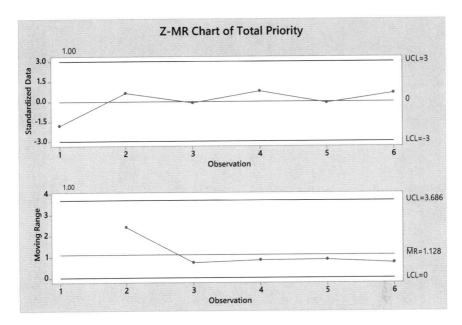

Fig. 3.16 The data demonstration in the domain of processing

Table 3.43 The extracted data from the selected fuzzy method in the inventory management

N	Name	Fuzzy method (FDEMATEL + FANP)	
		Target	Total priority
1	Redesign	1.00	0.27
2	Reduce	1.00	0.55
3	Reuse	1.00	1.42
4	Renew	1.00	1.59
5	Repair	1.00	0.61
6	Recycle	1.00	1.56

3.4.5 The Application of the Z-MR Chart for Obtained Data (Inventory Management)

Table 3.43 and Fig. 3.17 show the extracted data in the domain of inventory.

All data extracted from the selected fuzzy method in this domain regarding the considered target are under control.

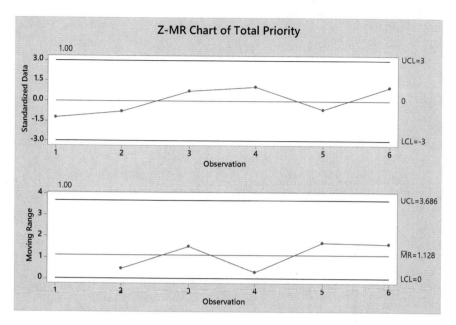

Fig. 3.17 The data demonstration in the domain of inventory

		Fuzzy method (FDEMATEL + FANP)	
N	Name	Target	Total priority
1	Redesign	1.00	0.35
2	Reduce	1.00	1.19
3	Reuse	1.00	1.56
4	Renew	1.00	1.10
5	Repair	1.00	0.44
6	Recycle	1.00	1.36

Table 3.44 The extracted data from the selected fuzzy method in the suppliers evaluation

3.4.6   The Application of the Z-MR Chart for Obtained Data (Suppliers Evaluation)

Table 3.44 and Fig. 3.18 show the extracted data in the domain of purchasing.

All data extracted from the selected fuzzy method in this domain regarding the considered target are under control.

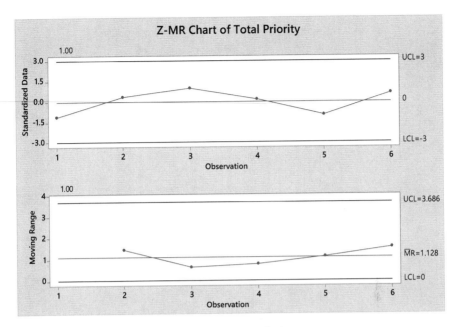

Fig. 3.18   The data demonstration in the domain of purchasing

			Fuzzy method (FDEMATEL + FANP)	
	N	Name	Target	Total priority
	1	Redesign	1.00	0.45
	2	Reduce	1.00	1.46
	3	Reuse	1.00	0.75
	4	Renew	1.00	1.53
	5	Repair	1.00	1.13
	6	Recycle	1.00	0.67

Table 3.45   The extracted data from the selected fuzzy method in the suppliers' quality

3.4.7   The Application of the Z-MR Chart for Obtained Data (Suppliers' Quality)

Table 3.45 and Fig. 3.19 show the extracted data in the domain of suppliers.

All data extracted from the selected fuzzy method in this domain regarding the considered target are under control.

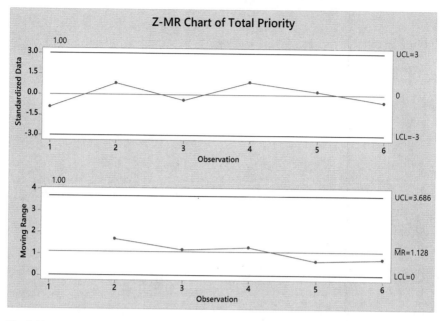

Fig. 3.19 The data demonstration in the domain of suppliers

Table 3.46 The extracted data from the selected fuzzy method in the location determination

| N | Name | Fuzzy method (FDEMATEL + FANP) | |
		Target	Total priority
1	Redesign	1.00	0.75
2	Reduce	1.00	0.91
3	Reuse	1.00	1.25
4	Renew	1.00	0.92
5	Repair	1.00	1.36
6	Recycle	1.00	0.81

3.4.8 The Application of the Z-MR Chart for Obtained Data (Location Determination)

Table 3.46 and Fig. 3.20 show the extracted data in the domain of location.

All data extracted from the selected fuzzy method in this domain regarding the considered target are under control.

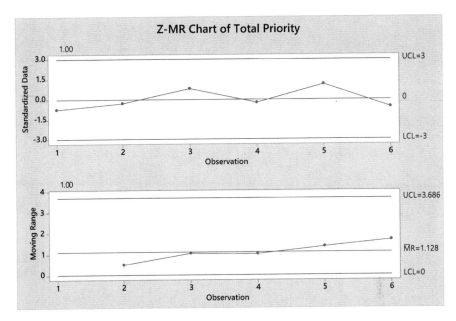

Fig. 3.20 The data demonstration in the domain of location

		Fuzzy method (FDEMATEL + FANP)	
Table 3.47 The extracted data from the selected fuzzy method in the movement and storage			
N	Name	Target	Total priority
1	Redesign	1.00	0.93
2	Reduce	1.00	0.67
3	Reuse	1.00	1.34
4	Renew	1.00	0.75
5	Repair	1.00	1.79
6	Recycle	1.00	0.52

3.4.9 The Application of the Z-MR Chart for Obtained Data (Movement and Storage)

Table 3.47 and Fig. 3.21 show the extracted data in the domain of logistics.

All data extracted from the selected fuzzy method in this domain regarding the considered target are under control.

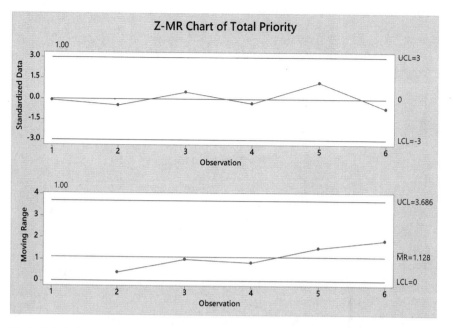

Fig. 3.21 The data demonstration in the domain of logistics

3.5   The Role of Process Capability Indices in Productivity and Sustainability

The role of process capability indices in managing productivity and sustainability has been studied and shown in previous studies (Rostamkhani and Karbasian 2020; Rostamkhani and Ramayah 2022). The management of the main goals (productivity and sustainability) needs to be assessed before and after the implementation of each approach/model. One of the best techniques for assessing the triple management processes is the process capability indices (PCI). The C_{pm} and C_{pmk} are defined based on the following formulas:

$$C_{pm} = \frac{USL - LSL}{6\sqrt{\sigma^2 + (\mu - T)^2}} \tag{3.2}$$

$$C_{pmk} = Min\left[\frac{USL - \mu}{3\sqrt{\sigma^2 + (\mu - T)^2}}, \frac{\mu - LSL}{3\sqrt{\sigma^2 + (\mu - T)^2}}\right] \tag{3.3}$$

The parameters of these formulas are defined as follows:

- USL is the upper specification limit and LSL is the lower specification limit.
- σ^2 is the variance, μ is the mean value, and T is the target value.

- The standard deviation in all calculations by default is $\sigma = 10$.
- The tolerance range in the field of indices is $\%100 \pm \%20$.

In general, the target value (T) in all processes is not equal to the mean value (μ) and the midpoint of technical specification limits (M). In this subsection, the relevant subprocesses and all data in the calculation of process capability indices for productivity and sustainability have been shown before implementing the proposed model in the book. The most important titles in this field are as follows:

1. C_{pmk} in Productivity Management Process (Mean Value of Productivity Data)
2. C_{pmk} in Sustainability Management Process (Mean Value of Sustainability Data)

The amount of process capability indices for productivity and sustainability will be shown after implementing the book's model in the elements of circular supply chain management in Sect. 3.5.3.

3.5.1 C_{pmk} in Productivity Management Process (Mean Value of Productivity Data)

Table 3.48 shows the C_{pmk} in the productivity management process in the circular thinking in a sample organization concerning the process context there.

The strongest C_{pm} and C_{pmk} belong to repair index and the weakest index C_{pm} and C_{pmk} belong to recycle index.

3.5.2 C_{pmk} in Sustainability Management Process (Mean Value of Sustainability Data)

Table 3.49 shows the C_{pmk} in the sustainability management process in the circular thinking in a sample organization concerning the process context there.

Table 3.48 Information in the productivity management process

Sub-process			Required indices Or Desired direction	U	Period	Target value (T)	Mean value (μ) Year	Process capability indices (PCI)	
N	Type	Name						C_{pm}	C_{pmk}
1	Main	Redesign	Increased Redesign	%	Year	100	92	0.52	0.31
2	Main	Reduce	Increased Reduce	%	Year	100	95	0.60	0.45
3	Leadership	Reuse	Increased Reuse	%	Year	100	94	0.57	0.40
4	Leadership	Renew	Decreased Renew	%	Year	100	93.5	0.56	0.38
5	Supportive	Repair	Decreased Repair	%	Year	100	98	0.65	0.59
6	Supportive	Recycle	Increased Recycle	%	Year	100	88	0.43	0.17

Table 3.49 Information in the sustainability management process

Sub-process			Required indices Or Desired direction	U	Period	Target value (T)	Mean value (μ) Year	Process capability indices (PCI)	
N	Type	Name						C_{pm}	C_{pmk}
1	Main	Redesign	Increased Redesign	%	Year	100	90	0.47	0.24
2	Main	Reduce	Increased Reduce	%	Year	100	92.5	0.53	0.33
3	Leadership	Reuse	Increased Reuse	%	Year	100	98	0.65	0.59
4	Leadership	Renew	Decreased Renew	%	Year	100	95.5	0.61	0.47
5	Supportive	Repair	Decreased Repair	%	Year	100	96	0.62	0.50
6	Supportive	Recycle	Increased Recycle	%	Year	100	85	0.37	0.09

Table 3.50 C_{pmk} average before and after the implementation of the approach (model)

The management of the main goals in Circular SCM	C_{pmk} average (before the implementation)	C_{pmk} average (after the implementation)	Increased percentage
Productivity Management Process	0.38	0.55	45%
Sustainability Management Process	0.37	0.58	57%

The strongest C_{pm} and C_{pmk} belong to reuse index and the weakest index C_{pm} and C_{pmk} belong to recycle index.

3.5.3 C_{pmk} Before and After the Implementation of the Proposed Model (Book's Model)

The relevant C_{pmk} average before and after implementing the proposed model for the elements of supply chain management by combining circular thinking in the same industry has been calculated and shown in Table 3.50.

Analysis:

As can be seen, the implementation of the proposed approach (model) can increase the C_{pmk} average for the double main goals by more than 40%. It is a strategic point in this research that significantly rises the C_{pmk} average for these vital goals simultaneously.

3.6   Conclusion

In this chapter, we understood how the design of experiments helps us to choose the best fuzzy technique and how the selected fuzzy technique has the required ability to implement the performance assessment of an organization after combining circular thinking in all elements of supply chain management step by step. In the next step, we have shown the impact of Z-MR control charts (SPC) to assess the output data from to be or not under control aspect. Moreover, in this chapter, we have shown the role of process capability indices in productivity and sustainability and we explained the impact of the proposed model on how to increase the process capability indices. Figure 3.22 demonstrates the all levels in this chapter in a summarized perspective for better understanding the readers and paying attention to the key outcomes.

Figure 3.23 thoroughly shows the main extracted results from Fig. 3.4 to Fig. 3.12.

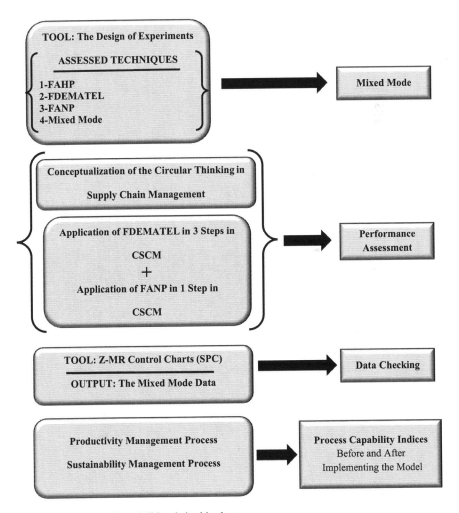

Fig. 3.22 Demonstration of all levels in this chapter

Customers	
Element of CT	Priority
Renew	1
Reduce	2
Reuse	3

Forecasting	
Element of CT	Priority
Reuse	1
Redesign	2
Renew	3

Designing	
Element of CT	Priority
Redesign	1
Reduce	2
Renew	3

Processing	
Element of CT	Priority
Renew	1
Reduce	2
Recycle	3

Inventory	
Element of CT	Priority
Renew	1
Recycle	2
Reuse	3

Purchasing	
Element of CT	Priority
Reuse	1
Recycle	2
Reduce	3

Suppliers	
Element of CT	Priority
Renew	1
Reduce	2
Repair	3

Location	
Element of CT	Priority
Repair	1
Reuse	2
Renew	3

Logistics	
Element of CT	Priority
Repair	1
Reuse	2
Redesign	3

Fig. 3.23 A summary of the main extracted results

References

Rostamkhani, R., & Karbasian, M. (2020). *Quality Engineering Techniques: An Innovative and Creative Process Model* (1st ed.). Taylor and Francis Group/CRC Press. https://doi.org/10.1201/9781003042037

Rostamkhani, R., & Ramayah, T. (2022). *A Quality Engineering Techniques Approach to Supply Chain Management* (1st ed.). Springer Nature. https://doi.org/10.1007/978-981-19-6837-2

Chapter 4
The Advantages and Achievements of the Book's Model

4.1 Introduction

In this chapter, the advantages and achievements of the model used in the book (combined fuzzy approach) compared to other models for measuring the circular supply chain performance of an organization will be explained. In other words, while referring to the previous methods and models and stating their disadvantages of them, the advantages of the combined fuzzy model used to measure the performance of the organization's circular supply network are described in this chapter in the first step. Indeed, although we will show that the previous methods have tried to cover the different aspects of the supply chain performance, it has not been introduced any fuzzy technique on the circular supply chain performance of an organization in a comprehensive format by an integrated approach (presented in Chap. 3). Then we will state the achievements of the extracted combined fuzzy model (as an output of the design of experiments) for the performance of the organization's circular supply chain management in this chapter. Figure 4.1 presents the previous methods in the different aspects of the supply chain performance of an organization in a summarized perspective. None of them has focused on the circular supply chain performance and achieving productivity and sustainability by fuzzy techniques through a comprehensive approach. This is the biggest competitive advantage of this book that is not visible in any book.

4.2 The Advantages of Applied Model (CFT+QET+PCI) in the Book

The previous viewpoints and performance measurement methods for the different aspects of supply chain performance at the organizational level can be designed and investigated by economists, engineers, managers, accountants, and mathematicians.

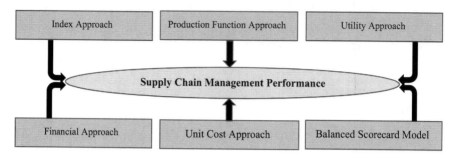

Fig. 4.1 The previous methods in the different aspects of performance measurement

The most traditional methods for assessing organizational performance are as follows:

(In some circumstances, these methods can be indirectly applied to the circular supply chain management performance in an effective format by scholars.)

4.2.1   Index Approach

The mentioned method is more desired by economists and engineers. Kendrick-Kremer, Craig-Harris Model, Hines' Model, American Productivity and Quality Center (APQC), and Sumanth models are among the models presented in this method.

4.2.2   Production Function Approach

The models of production functions used by economists such as the Cobb–Douglas model, production functions with constant elasticity of substitution (CES), and transcendental production functions, Spillman, Teaf, and Translog are based on different experimental observations. It should be noted that they start with the initial assumptions regarding environmental elements.

4.2.3   Utility Approach

The mentioned method has attracted the attention of engineers that the studies of Stewart, Hershauer, & Ruch are important in this field.

4.2.4   Financial Approach

This method is used by managers and accountants. Types of financial ratios, Gold and Aggarwal models, Quick Productivity Appraisal Approach and Value Added, Lawler's method, and Mao's method are included in this category. Due to its nature, this technique is one of the few techniques that can directly measure the performance of supply chain management elements. In other words, this technique can be used in any situation to supply chain network data.

4.2.5   Unit Cost Approach

The mentioned method is more interesting to accountants and managers. The unit cost analysis is based on the production hall, department, and product based on this method. Adam's research has been presented based on the QPR (Quality Productivity Ratio) model to express quality changes in an organization using this method.

4.2.6   The Balanced Scorecard Model

In the mentioned method, the vision and strategy form the central core of this evaluation method. By using this framework, the balanced evaluation method transforms the vision and strategy of the organization into general goals, related metrics, quantitative goals, and implementation plans and initiatives to measure the performance of the supply chain network.

4.2.7   Data Envelopment Analysis

The mentioned method is based on a series of optimizations using linear programming, which is also called a non-parametric method. In this method, the efficient frontier curve is created from a series of points determined by linear programming. Another important feature of DEA evaluation is the combined evaluation of a set of factors. In DEA models, input and output factors are evaluated together and there is no limitation of one input or one output. Although the data coverage analysis was initially presented to evaluate the decision-making units, the extensive capabilities of its models have introduced this method as one of the most widely used methods, especially for evaluating the performance of supply chain network data in organizations.

4.2.8   The Other Types of Studies Related to the Circular SCM Performance

In other situations, some advanced studies can be directly applied to the circular supply chain management performance by scholars. The main core of these studies is based on focusing on the circular supply chain management performance. Although these studies are more advanced compared to the previous methods, there is no clue about applying fuzzy techniques and quality engineering techniques as facilitator tools in the framework of the circular supply chain management performance in them. Table 4.1 shows these advanced studies.

None of the above studies has a comprehensive approach to the performance measurement of circular supply chain management. The combination of statistical techniques as a facilitating tool and fuzzy techniques as a main tool is completely new and can increase the ability to achieve productivity and sustainability more than previous research.

4.2.9   The Advantages of Applying the Book's Model (CFT +QET+PCI)

Before presenting this book, as can be seen, several methods and studies have been proposed to evaluate the performance of organizations in general and elements of circular supply chain management in particular. Traditional methods focus on financial aspects and advanced methods focus on non-financial aspects. Meanwhile, only focusing on financial aspects is not enough to deal with changes in the business environment especially related to the circular supply chain management performance. Of course, new methods and techniques, such as the balanced scorecard and the data envelopment analysis, while they try to compensate for the current

Table 4.1 Some advanced studies about the circular supply chain management performance

Title	Author(s)	Date
Strategic framework towards measuring a circular supply chain management	Jain, S., Jain, N.K. and Metri, B.	2018
Analyzing the circular supply chain management performance measurement framework: the modified balanced scorecard technique	Saroha, M., Garg D., Luthra, S.	2022
A proposed circular-SCOR model for supply chain performance measurement in manufacturing industry during COVID-19	Ozbiltekin-Pala, M., Koçak, A. and Kazancoglu, Y.	2023
Supplier performance and selection from sustainable supply chain performance perspective	Meena, P.L., Katiyar, R. and Kumar, G.	2022
Circular supply chain implementation performance measurement framework: a comparative case analysis	Lahan, S., Kant, R., Shankar R., Patil, S.K.	2023

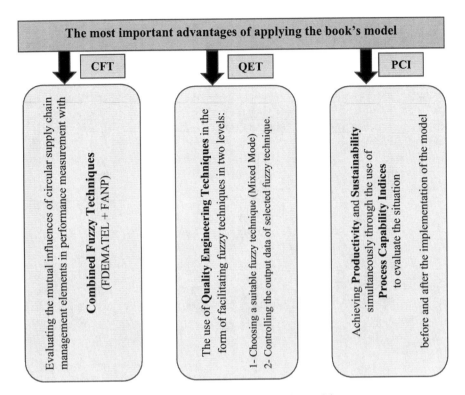

Fig. 4.2   The most important advantages of applying the book's model

disadvantages, but they do not succeed completely. Indeed, it seems that by combining the two balanced scorecard models and data envelopment analysis, not only the goals and strategic indicators of the organization can be evaluated in a longitudinal process, but also the performance comparison between different units and the evaluation of the organization's performance at certain points can be done at the same time. In addition, the combination of these two techniques can perform better in evaluating the performance of circular supply network data. But there are still weaknesses that even adding European Foundation for Quality Management (EFQM) to these techniques cannot compensate for. The biggest competitive advantage of the book's model is to consider the mutual effects of circular supply chain management elements on each other. It means that a comprehensive evaluation of the performance of circular supply chain management is possible only in fuzzy studies and the use of fuzzy techniques. Moreover, the use of quality engineering techniques can guarantee the assessment conditions as the facilitating tools and provide our goal to reach productivity and sustainability. Figure 4.2 shows the most important advantages of applying the book's model.

4.3 The Achievements of Applied Model (CFT+QET+PCI) in the Book

In the competitive and intensive industrial world, only organizations can survive and compete that are continuously evaluating and improving their performance, especially in the discussion of supply chain network management, and according to advanced concepts such as circular concepts. They should reach an acceptable level to have performance to achieve productivity and sustainability. In this subsection, we will discuss the achievements of the three sections of the book's model and show what exactly each section will bring to the industries.

4.3.1 The Achievements of Applying Combined Fuzzy Techniques (CFT)

Figure 4.3 shows the exact achievements of applying combined fuzzy techniques (FDEMATEL + FANP) in evaluating the performance of circular supply chain management.

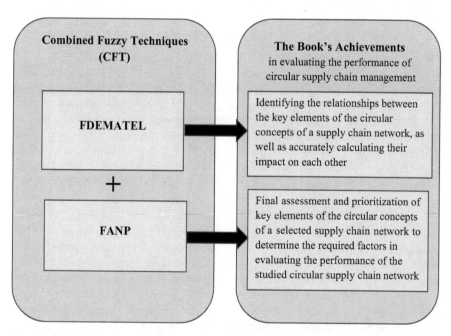

Fig. 4.3 The achievements of applying combined fuzzy techniques (FDEMATEL + ANP)

4.3.2   The Achievements of Applying Quality Engineering Techniques (QET)

Figure 4.4 shows the exact achievements of applying quality engineering techniques (DOE + SPC) in introducing the facilitating role of key quality engineering techniques such as DOE and SPC (Z-MR Charts) for fuzzy techniques.

4.3.3   The Achievements of Applying Process Capability Indices (PCI)

One of the best achievements of applying the book's model is to introduce the implementation of process capability indices to the effective management of productivity and sustainability processes related to the circular supply chain management performance. The vital achievements of applying it are as follows:

1. The introduction of subprocesses in productivity and sustainability concerning the circular concepts and the calculations of C_{pmk} can create more understanding of the proposed model in the book.
2. The demonstration of the C_{pmk} average before and after implementing the proposed model for the elements of circular supply chain management can show the impact of the implementation of the model.

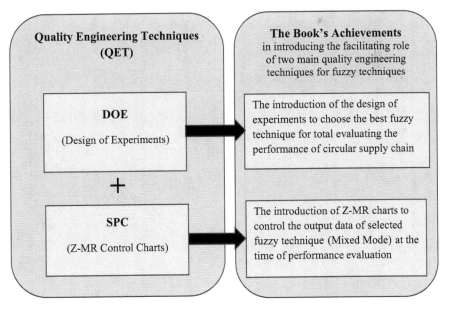

Fig. 4.4   The achievements of applying quality engineering techniques (DOE + SPC)

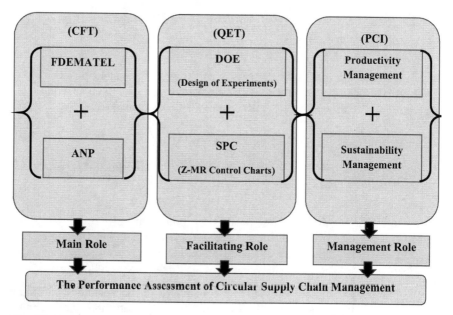

Fig. 4.5 A summary of advantages and achievements of the book's model

4.4   Conclusion

In this chapter, we reviewed the previous models or approaches in assessing supply chain management for the direct or indirect assessment related to the performance of circular supply chain management. Also, the proposed model of the book has been presented in the three sections as follows:

1. Combined Fuzzy Techniques (CFT)
2. Quality Engineering Techniques (QET)
3. Process Capability Indices (PCI)

In the next stage, the most important advantages and achievements have been explained in detail. Figure 4.5 shows a summary of this chapter in a whole perspective.

References

Jain, S., Jain, N. K., & Metri, B. (2018). Strategic framework towards measuring a circular supply chain management. *Benchmarking: An International Journal, 25*(8), 3238–3252. https://doi. org/10.1108/BIJ-11-2017-0304

Lahan, S., Kant, R., Shankar, R., & Patil, S. K. (2023). Circular supply chain implementation performance measurement framework: a comparative case analysis. *Production Planning & Control*. Ahead-of-print. https://doi.org/10.1080/09537287.2023.2180684

Meena, P. L., Katiyar, R., & Kumar, G. (2022). Supplier performance and selection from sustainable supply chain performance perspective. *International Journal of Productivity and Performance Management*. Ahead-of-print. https://doi.org/10.1108/IJPPM-01-2022-0024

Ozbiltekin-Pala, M., Koçak, A., & Kazancoglu, Y. (2023). A proposed circular-SCOR model for supply chain performance measurement in manufacturing industry during COVID-19. *International Journal of Quality & Reliability Management, 40*(5), 1203–1232. https://doi.org/10.1108/IJQRM-03-2022-0101

Saroha, M., Garg, D., & Luthra, S. (2022). Analyzing the circular supply chain management performance measurement framework: the modified balanced scorecard technique. *International Journal of System Assurance Engineering and Management, 13*(2), 951–960. https://doi.org/10.1007/s13198-021-01482-4

Chapter 5
The Perspective of Future Horizon for the Advanced Studies

5.1 Introduction

One of the most important topics related to the title of this book is the review and proposal of the future path for measuring circular supply chain management performance with advanced mathematical functions and differential equations. The role of mathematical functions and differential equations will be extremely vital for the development of the subject discussed in the book, especially for economic theorists in this field. Figure 5.1 shows these two vital perspectives for future horizons.

5.2 How to Use the Advanced Mathematical Functions to Extend Measurement

Tables 5.1, 5.2, 5.3, 5.4, 5.5, 5.6, 5.7, 5.8 and 5.9 show the most important factors of circular thinking in the different elements of supply chain management. Table 5.10 introduces the independent variables in the measurement of the circular supply chain management performance in the numerical application of the book (Chap. 3).

Although in recent years, there are individually valuable studies related to the mathematical modeling of the supply chain management elements (Easters 2014; Forozandeh et al. 2017; Mohammad and Kazemipoor 2020; Rajabzadeh Gatari et al. 2021; Alkahtani 2022; Bahrampour et al. 2022), there is no reported study about the formulation of circular thinking elements in the supply chain elements to measure performance to achieve productivity and sustainability. But creating advanced linear and non-linear mathematical functions to extend measurement can be closer to the study related to a model for supply chain design considering the cost of quality (Castillo-Villar et al. 2012). The important notes are to consider the appropriate target functions in each element of supply chain management concerning the restrictions in the circular elements. Figure 5.2 shows the total framework of

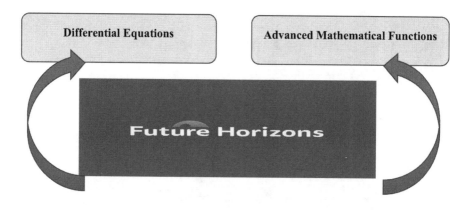

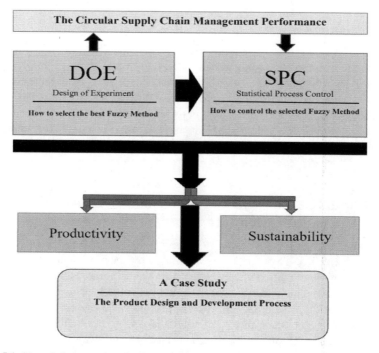

Fig. 5.1 Two vital perspectives for future horizons

Table 5.1 The most important factors of circular elements in the customers section of SCM

Customers		
Degree of influence (1)	Degree of influence (2)	Degree of influence (3)
Renew	Reduce	Reuse
60.7	14.8	−2.9

Table 5.2 The most important factors of circular elements in the forecasting section of SCM

Forecasting		
Degree of influence (1)	Degree of influence (2)	Degree of influence (3)
Reuse	Redesign	Renew
28.9	22.6	13.0

Table 5.3 The most important factors of circular elements in the designing section of SCM

Designing		
Degree of influence (1)	Degree of influence (2)	Degree of influence (3)
Redesign	Reduce	Renew
43.6	34.3	27.9

Table 5.4 The most important factors of circular elements in the processing section of SCM

Processing		
Degree of influence (1)	Degree of influence (2)	Degree of influence (3)
Renew	Reduce	Recycle
24.6	20.6	20.1

Table 5.5 The most important factors of circular elements in the inventory section of SCM

Inventory		
Degree of influence (1)	Degree of influence (2)	Degree of influence (3)
Renew	Recycle	Reuse
59.1	56.4	41.7

Table 5.6 The most important factors of circular elements in the purchasing section of SCM

Purchasing		
Degree of influence (1)	Degree of influence (2)	Degree of influence (3)
Reuse	Recycle	Reduce
55.8	35.6	18.9

Table 5.7 The most important factors of circular elements in the suppliers section of SCM

Suppliers		
Degree of influence (1)	Degree of influence (2)	Degree of influence (3)
Renew	Reduce	Repair
52.6	46.1	13.3

mathematical modeling applicable in this domain. Indeed, this approach can reach comprehensive mathematical modeling in the relevant research.

The general target function in a supply chain network for applying the circular thinking elements can be presented as follows:

Table 5.8 The most important factors of circular elements in the location section of SCM

Location		
Degree of influence (1)	Degree of influence (2)	Degree of influence (3)
Repair	Reuse	Renew
35.8	25.0	−8.1

Table 5.9 The most important factors of circular elements in the logistics section of SCM

Logistics		
Degree of influence (1)	Degree of influence (2)	Degree of influence (3)
Repair	Reuse	Redesign
78.7	34	−7

Table 5.10 The introduction of independent variables in the measurement of the CSCM performance

The circular elements X_{ij}	The number of repetitions in SCM Independent variable (j=1)	The average of influence Independent variable (j=2)
Redesign i=1	X_{11}	X_{12}
	(In the book's numerical application equals 3)	(In the book's numerical application equals 19.73)
Reduce i=2	X_{21}	X_{22}
	(In the book's numerical application equals 5)	(In the book's numerical application equals 26.94)
Reuse i=3	X_{31}	X_{32}
	(In the book's numerical application equals 6)	(In the book's numerical application equals 30.42)
Renew i=4	X_{41}	X_{42}
	(In the book's numerical application equals 7)	(In the book's numerical application equals 32.83)
Repair i=5	X_{51}	X_{52}
	(In the book's numerical application equals 3)	(In the book's numerical application equals 42.6)
Recycle i=6	X_{61}	X_{62}
	(In the book's numerical application equals 3)	(In the book's numerical application equals 37.37)

$$\text{MAX } Q = \sum_{k=1}^{9} W_k \Phi_k \qquad (5.1)$$

Q: The overall productivity and sustainability level achieved by the CSCM

W_k: The proportion of adaptability to circular thinking elements from the first to sixth

Φ_k: The total proportion of adaptability in achieving productivity and sustainability

Target Functions (Supply Chain Management) Φ_1 to Φ_9	Restrictions (Circular Elements) R_1 to R_9
Customer (Φ_1)	$R_1 = \sum_{i=1}^{i=6} \sum_{j=1}^{j=2} C_{ij1} X_{ij1}$
Forecasting (Φ_2)	$R_2 = \sum_{i=1}^{i=6} \sum_{j=1}^{j=2} C_{ij2} X_{ij2}$
Designing (Φ_3)	$R_3 = \sum_{i=1}^{i=6} \sum_{j=1}^{j=2} C_{ij3} X_{ij3}$
Processing (Φ_4)	$R_4 = \sum_{i=1}^{i=6} \sum_{j=1}^{j=2} C_{ij4} X_{ij4}$
Inventory (Φ_5)	$R_5 = \sum_{i=1}^{i=6} \sum_{j=1}^{j=2} C_{ij5} X_{ij5}$
Purchasing (Φ_6)	$R_6 = \sum_{i=1}^{i=6} \sum_{j=1}^{j=2} C_{ij6} X_{ij6}$
Suppliers (Φ_7)	$R_7 = \sum_{i=1}^{i=6} \sum_{j=1}^{j=2} C_{ij7} X_{ij7}$
Location (Φ_8)	$R_8 = \sum_{i=1}^{i=6} \sum_{j=1}^{j=2} C_{ij8} X_{ij8}$
Logistics (Φ_9)	$R_9 = \sum_{i=1}^{i=6} \sum_{j=1}^{j=2} C_{ij9} X_{ij9}$

Fig. 5.2   The suggested framework for modeling using the appropriate mathematical functions

$$R_k = \sum_{i=1}^{i=6} \sum_{j=1}^{j=2} \sum_{k=1}^{k=9} C_{ijk} X_{ijk} \qquad \text{(S.T)}$$

By considering the proposed target function, in addition to the definition of circular elements as the first constraint, other possible constraints related to cost, time, and other consumption resources can be defined in this matter as a multi-objective problem. Referring to the assumptions in the valuable study done by other scholars can help the interested readers to reach a comprehensive format in this domain (Chin et al. 2018). Also, we can solve them by linear or non-linear approach.

5.3 The Suggestions for the New Directions Utilizing the Differential Equations

Differential equations are an important mathematical concept used to model processes in many disciplines (Sijmkens et al. 2022). Differential equations are great for modeling situations where there is a continually changing population or value. The theory of differential equations has become an essential tool of management analysis such as the continual changing of the supply chain management elements affected by circular thinking particularly since the computer has become commonly available. One of the best studies related to the application of ordinary and partial differential equations in supply chain networks are the fourth and fifth chapters of a book written by D'Apice et al. (2010). This book has presented beneficial examples of these applications. But there is no sufficient information on the application of differential equations in the combination of the supply chain and the circular elements in this book or other publications.

If a reference differential equation has been introduced by us that can cover the overall productivity and sustainability level achieved by the circular supply chain management influenced by the circular thinking elements, no doubt, no doubt, the suggested differential equations can be as follows:

$$Q = f(X_1, \ldots, X_9) \tag{5.2}$$

f: The defined function to show the relationship between Q and the SCM elements
Q(Independent Variable): The overall productivity and sustainability level achieved by the CSCM
$X_1, \ldots, X_9$: The variables from the first element (customer) to the ninth element (logistics)

$$Q = X_1 \frac{\partial Q}{\partial X_1} + \ldots + X_9 \frac{\partial Q}{\partial X_9} \tag{5.3}$$

$$X = g(x_1, \ldots, x_6) \tag{5.4}$$

g: The defined function to show the relationship between X and the circular elements
X(Independent Variable): The overall supply chain management influenced by the circular elements

$x_1, \ldots, x_6$: The variables from the first element (redesign) to the sixth element (recycle)

$$X = x_1 \frac{\partial X}{\partial x_1} + \ldots + x_6 \frac{\partial X}{\partial x_6} \qquad (5.5)$$

5.4 Conclusion

The main goal of this chapter is to discuss future horizons by focusing on advanced mathematical functions and differential equations. This chapter can be used by theoretical researchers in universities that intend to extend the main theoretical domains of circular supply chain management. Considering multi-objective programming and adding variables that cover all issues related to cost and time are vital to extending the measurement of performance in circular supply chain management. Moreover, the definition of the general framework of differential equations to explain the relationship between the supply chain management elements and the circular thinking elements can be to establish the main basis of each theoretical study in the future. When we want to consider the appropriate independent variables in differential equations related to the performance measurement of circular supply chain management, it is better to have all independent variables related to economic, environmental, and social performance. It is vital to reach productivity and sustainability and anybody knows that this way can completely go through economic, environmental, and social issues.

References

Alkahtani, M. (2022). Mathematical modelling of inventory and process outsourcing for optimization of supply chain management. *Mathematics, 10*(7), 1142. https://doi.org/10.3390/math10071142

Bahrampour, P., Najafi, S. E., Lotfi, F. H., & Edalatpanah, S. H. (2022). Development of scenario-based mathematical model for sustainable closed loop supply chain considering reliability of direct logistics elements. *Journal of Quality Engineering and Production Optimization, 7*(2). https://doi.org/10.22070/jqepo.2022.15643.1219

Castillo-Villar, K. K., Smith, N. R., & Simonton, J. L. (2012). A model for supply chain design considering the cost of quality. *Applied Mathematical Modelling, 36*(12), 5920–5935. ISSN 0307-904X. https://doi.org/10.1016/j.apm.2012.01.046

Chin, Y. S., Seow, H. V., Lee, L. S., & Kumar, R. (2018). Fuzzy Mathematical model for solving supply chain problem. *Journal of Computer and Communications, 6*(9). https://doi.org/10.4236/jcc.2018.69007

D'Apice, C., Gottlich, S., Herty, M., & Piccoli, B. (2010). *Modeling, Simulation, and Optimization of Supply Chains*. Book Series Name: Other Titles in Applied Mathematics., ISBN: 978-0-89871-700-6, eISBN: 978-0-89871-760-0. https://doi.org/10.1137/1.9780898717600

Easters, D. J. (2014). *Mathematical supply-chain modelling: product analysis of cost and time.* *Journal of Physics: Conference Series, 2nd International Conference on Mathematical Modeling in Physical Sciences 2013, Vol 490.* Prague, Czech Republic. https://doi.org/10.1088/1742-6596/490/1/012041

Forozandeh, M., Teimoury, E., & Makui, A. (2017). A mathematical formulation of time-cost and reliability optimization for supply chain management. *RAIRO-Operations Research, 53*(4), 1385–1406. https://doi.org/10.1051/ro/2018068

Mohammad, P., & Kazemipoor, H. (2020). An integrated multi-objective mathematical model to select suppliers in green supply chains. *International Journal of Research in Industrial Engineering, 9*(3), 216–234. https://doi.org/10.22105/riej.2020.262937.1173

Rajabzadeh Gatari, A., Amini, M., Azar, A., & Kolyaei, M. (2021). Design of integrated mathematical model for closed-loop supply chain. *Management Research in Iran, 20*(1), 1–32. 20.1001.1.2322200.1395.20.1.1.6

Sijmkens, E., Scheerlinck, N., Cock, M. D., & Deprez, J. (2022). Benefits of using context while teaching differential equations. *International Journal of Mathematical Education in Science and Technology.* https://doi.org/10.1080/0020739X.2022.2039412

Chapter 6
A Case Study in the Product Design and Development Process

6.1 Introduction

In the final chapter of the book, we describe a case study of the implementation of circular lean and agility in supply chain networks for the process of product design and development. This is an effective way to get readers to practically apply the book's model to their implementation of learned content in a real-world situation. In general, the process of product design and development is at the beginning of supply chain management or supply chain networks. Although the concepts of circular lean and circular agile in supply chains are used simultaneously, it should be noted that in the product design section, priority is given to the circular lean concept, and in the product development section, the priority is to the circular agile concept. Of course, we will not ignore the concepts of circular agility in the design process, as well as the concepts of circular lean in the product development process. In other words, in the presented case study, in the design of the lean and agile circular thinking development model, the model has been designed to lead to the implementation of the principles of circular lean thinking in the product design process. Simultaneously, in the field of product development, the capability of circular agility has to be available. For this reason, in the implementation of the model of circular lean-agility concepts, it has been tried to follow the principles of simultaneous development. Measuring the performance of a circular supply chain management system based on circular lean-agile thinking in product design and development process will be the final goal.

6.2 A Summary of Case Study Background

So far, many studies have been presented in the field of applying the concepts of circular lean thinking or circular agile thinking in the process of product design and development; these studies did not have an integrated approach and each emphasized one of the components of the process. For example, some studies have focused on identifying waste in the product design and development process, some have emphasized enablers, some have focused on human resource management in the product design and development process, and some have focused on other important areas of lean or agile thinking. In this case study, an attempt has been made to explain the combined model of elements of supply chain management with circular thinking in a lean and agile application perspective for the process of product design and development for all readers who are interested in implementing the proposed model. One of the most important weaknesses of previous studies in the simultaneous application of the concepts of circular lean and circular agility in supply chain networks has been that there has always been a tendency towards one of these concepts and the other concept has been neglected.

6.3   The Definition of Circular Lean and Agility in Supply Chain Networks

- *Circular Lean:* The concept of circular lean means the simultaneous combination of two frameworks: 1—Circular Concepts, 2—Lean Concepts (CL_1, CL_2, CL_3, CL_4) (Fig. 6.1)

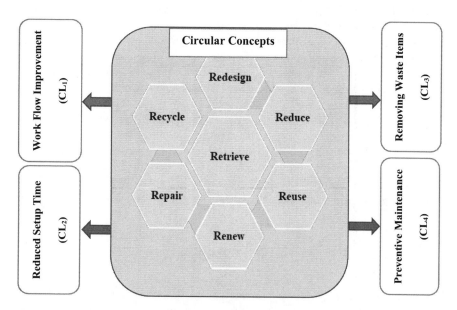

Fig. 6.1   The combination of two frameworks in circular lean

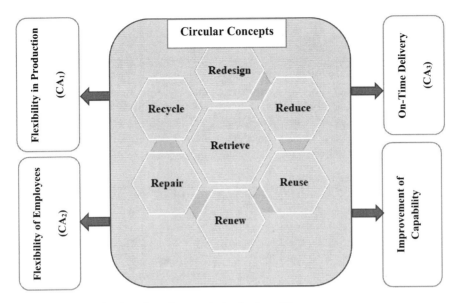

Fig. 6.2   The combination of two frameworks in circular agility

- *Circular Agility:* The concept of circular agility means the simultaneous combination of two frameworks: 1—Circular Concepts, 2—Agile Concepts (CA_1, CA_2, CA_3, CA_4) (Fig. 6.2).

6.4   The Circular Lean and Agility in the Product Design and Development

The most important steps required in the product design and development process are as follows:

Step 1: The product design and development process start from the value statement or product mission statement and will continue until the preparation of the product plan for mass production. The design process will continue until the "acceptance of the product design accurately by the production stakeholders." This acceptance depends on a sufficient and accurate knowledge of "what should be produced," "how to produce it," and "production organizational requirements" regarding the details and the whole system.

Step 2: One of the most basic approaches required in the product design process is determining the types of value-added activities that are divided into three categories (product design):

1. Value-added activities without which the product will be defective.
2. Non-value-added activities whose elimination will lead to improved production performance.

3. Activities without necessary added value that do not create added value, but their existence is necessary.

Activities without necessary added value (third category) are usually activities that are created due to the existence of bureaucracies and legal regulations. One of the considerations of the above category is that many times the activities of the second and third category hide themselves in activities with added value and great. This issue requires that the way of performing activities be analyzed in detail. Many experts divided the non-added value performance in the product design and development process into seven general categories as follows:

1. Unnecessary transportation
2. Maintenance of information more than required
3. Excessive information (production of unnecessary information)
4. Unnecessary movements (movement of people to obtain information)
5. Processing more than necessary (unnecessary work to produce output)
6. Waiting (delay to receive information, data, input, verification, distribution, or...)
7. Defects or failure (weakness in the quality of information and the need to rework to produce it)

Step 3: One of the most basic approaches required in the product development process is to respond to changes and threats in the best way and in the shortest time, as well as discover opportunities and provide successful solutions to exploit them. To realize a prosperous and competitive design and development, the experts considered the following points (product development):

1. Evaluation of the organization's capability: Organizations must first provide evaluation tools to determine the risks and threats around the organization and their strengths in responding to them. This assessment helps them to understand what activities they are ready to do.
2. Assurance of responsibilities: For the successful implementation of design and development processes, the organization must prepare the necessary plans to determine the roles (designation of executive directors, department managers, teams, etc.) for the implementation of their plans.
3. Understanding current processes: Organizations should carefully review their current processes and ensure they are properly understood before attempting to change them. Then we should carefully identify the problems and try to do well in the same field.

6.4.1   The Application of Circular Lean in Added Value Activities (Step 2)

Figure 6.3 shows the application of circular lean in added value activities (product design) as follows:

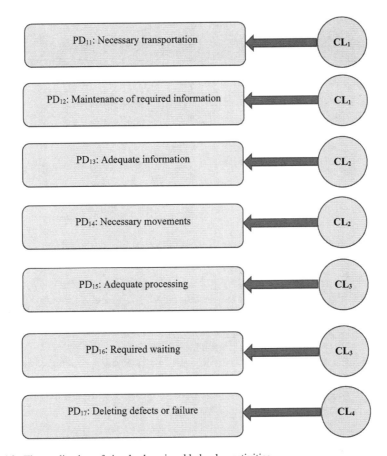

Fig. 6.3 The application of circular lean in added value activities

6.4.2   The Application of Circular Agility to Respond to Changes (Step 3)

Figure 6.4 shows the application of circular agility to respond to changes (product development) as follows:

6.4.3   General Weaknesses in the Product Design and Development Process

The most important reasons for weakness in product design and development process that circular lean and circular agility in supply chain networks or supply

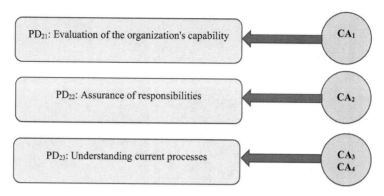

Fig. 6.4 The application of circular agility to respond to changes

chain management can be effective in addition to those mentioned above in correcting are generally:

− Poor schedule management
− Non-optimal use of expert human resources
− Existence of bureaucracy and excessive conservatism
− Failure to use appropriate databases for the design process
− Poor planning and management of the product development process
− Not using a tension system and using a pressure system in the design process
− Excessive emphasis on design points, lack of sufficient explanations for designs
− Failure to apply knowledge management in product design and development projects and starting each program regardless of the experiences gained from past programs
− The lack of necessary and sufficient coordination and the existence of poor communication between team members, especially between different units or between the manufacturer and different parts of the supply chain elements.

6.4.4   An Overview of the Product Design and Development Process

An overview of the product design and development process can be seen in Fig. 6.5 and the role that circular lean and circular agility can play there. The product design and development process starts by defining the value and schedule of activities, which is set and presented with the help of the value stream map, and the process ends with the production of the prototype of the product that can be presented to the final customers. Pay attention to the role impact of single to single of all components in this case study as follows:

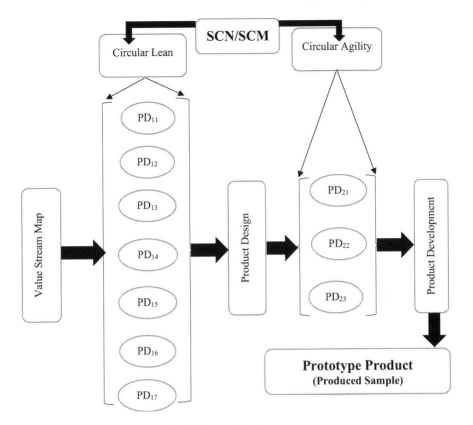

Fig. 6.5 An overview of the product design and development process

1. Value stream map
2. Product design process
3. Product development process
4. Circular lean extracted from SCN/SCM
5. Circular agility extracted from SCN/SCM
6. Prototype product (sample product) as a final output in this case study

In general, the product design and development process should begin with the precise definition of expectations from the desired product through the product value statement. The statement of product value makes it possible to avoid errors in the product production process and to meet the needs of customers while reducing the production cost and taking maximum care of the quality. Therefore, in the definition of the value statement, all required practical and operational definitions of the product should be provided. The time to complete the product design and development process is one of the key issues in explaining the value statement. In fact, as in the production line, determining the time to produce products is very important, in the model proposed in this case study, determining the time to complete the design and development process of the product is also very important. In determining this time, things such as the time customers need for products and competitive conditions should be considered. Factors such as the cost of the program, technology changes, the number of changes in the use of human resources, and the cash flow of the product design and development process are other considerations that can play a role in determining the design and development time of the product in addition to the main considerations. Therefore, it can also be claimed at this stage that the concepts of circular lean and circular agility are very useful. In summary, regarding the role of product design and development in a circular supply chain network or circular supply chain management, we reached the practical conclusion that the concepts of circular lean in the product design section and circular agility concepts in the product development section can create the best results. One of the most vital issues in the success of the implementation of lean-agile circular thinking based on the elements of supply chain management or supply chain networks in the product design and development process is the necessity of training participants in the process of product design and development to learn the principles of lean-agile circular thinking. These people should understand the importance of the value stream map and be able to handle the implementation of circular lean in the design section and the implementation of circular agility in the development section. The roles of designer experts, chief designer manager, and assistant chief designer manager should be trained in a clear and precise manner. All people must have the necessary training to understand the importance and role of communication and coordination required and the relevant tools for the success of the process.

6.4.5 Operational Implementation of a Design and Development Process

Action plans in a product design and development process that can be considered simultaneously with the relevant circular lean or circular agility in supply chain networks (supply chain management) approaches are as follows:

- Planning for employee training
- Automation and management of information flow
- Expanding the culture of continuous improvement
- Expansion of teamwork in process implementation
- Documentation and knowledge management in the process
- Designing coordination mechanisms between internal stakeholders
- Designing participation mechanisms for customers and final consumers
- Planning and designing tools for identifying non-value-creating activities
- Clarifying and explaining the duties of the persons involved in the process
- Designing the value stream and scheduling the execution of the process activities
- Designing and developing coordination tools with suppliers of parts and raw materials

- Identifying process uncertainties and necessary mechanisms to be flexible in the required activities
- Designing formal and informal communication processes for coordination between different activities
- Development and use of computer software to carry out product design and development process activities

6.5 Defining Indicators in the Circular Lean-Agility of SCN/SCM[1] in PDDP[2]

In planning for product design and development, managers and planners must gain complete knowledge of the current situation and prioritize their plans based on this knowledge. In the development of circular lean-agile thinking in a supply chain network in the product design and development process, it is essential to know the current state of the product design and development process and its most important weaknesses and strengths. On the other hand, managers need tools that can evaluate the progress of programs through these tools. One of these tools is the use of effective indicators in the covering of the planned product design and development process. Naturally, techniques such as the design of experiments or statistical control of the process will have their facilitating role and will help design managers effectively in choosing these tools or controlling the outputted data. In this chapter, according to the model explained in the book, as well as using balanced scorecard tools and the network analysis process, a model has been designed and presented to identify important indicators in evaluating the current situation and the communication of each of these indicators with the circular lean-agility of supply chain networks. After determining the key indicators and factors for the establishment of circular lean-agile thinking in the supply chain network, these criteria have been tried to be categorized in the framework of a balanced scorecard. This classification helps managers to implement circular lean-agile thinking regularly and systematically to plan in line with moving towards the design and development process of products in the supply chain network. Finally, to rank the required factors and criteria, the network analysis process has been used. The advantage of the network analysis process over other ranking methods is that all the connections between the selected indicators can be identified. In the balanced scorecard, the indicators of the financial aspect are the indicators that operate in a delayed manner, in other words, they are the indicators that are affected by the planning results in other balanced scorecard elements. Therefore, in this study, the defined indicators of the customer, the defined indicators of process, and the defined indicators of knowledge and human capital management are considered the main indicators that planners should

---

[1] Supply Chain Networks/Supply Chain Management.
[2] Product Design and Development Process.

evaluate the current state of the product design and development process. Each of the elements of the balanced scorecard model is considered a node of the network analysis process nodes, and each node is composed of several indicators or factors, and the final goal is to determine the weight and ranking of these factors and indicators. In defining the indicators, the following considerations were considered:

- Due to the necessity of a small number of indicators in the network analysis process (to prevent inconsistencies in pairwise comparisons), it was tried to eliminate the indicators that have a lot in common in the coverage of the evaluated features from the point of view of experts or have a cause-and-effect relationship.
- According to the nature of the product design and development process, some circular purity and agility indicators were tried to be redefined for this process.
- The most important output of the product design and development process will be the information about the product design to be presented to the market and customers. Also, the most important users of the output of the process are domestic production units and suppliers of parts and raw materials, and in the last stage, the final customers of the products.
- Based on the framework of the balanced scorecard, the defined indicators are placed in three levels, each level eventually forms one of the nodes of the network analysis process, and each of the indicators in each level is considered as one of the elements in the corresponding node.

All required indicators that simultaneously consider both circular lean-agility thinking in the design and development stages of the product and use comprehensive frameworks such as the four indicators of the balanced scorecard model have been applied in the chapter model.

In this chapter, the power of the FANP method is used to show the relationships between indicators when using the balanced scorecard method. The FANP method shapes the relationships between the indicators and prioritizes the indicators based on four main perspectives in the balanced scorecard. Also, using only FANP without entering BSC causes some vital determining factors to remain far away, which shows the necessity of using these two methods together. In addition, to accurately determine the relationships between the indicators and the degree of direct and indirect influence of the indicators on each other, as well as to investigate the influence of the four aspects of the balanced evaluation system on each other, the FDEMATEL decision-making system is used to strengthen the FANP method. Pay attention to the four main elements of BSC model as follows:

1. Financial
2. Customer
3. Internal business process
4. Organizational capacity process

Figure 6.6 shows the mentioned framework in a comprehensive perspective as follows:

BSC	Indicators in the Product Design and Development Process (Combined Mode)	Circular Lean-Agility (SCM/SCN)
Financial	Reducing the Number of Extra Employees	$CL_1 + CA_4$
	The Ratio of Liabilities to Assets	$CL_1 + CA_4$
	Commercial Exploitation	$CL_1 + CA_4$
	Reduction in Costs	$CL_1 + CA_4$
	Cost Forecast	$CL_1 + CA_4$
	Profitability	$CL_1 + CA_4$
	Cash Flow	$CL_1 + CA_4$
	Efficiency	$CL_1 + CA_4$
Customer	Higher Quality and Lower Price	$CL_1 + CL_2 + CA_3 + CA_4$
	Product or Service Availability	CA_3
	Flexible Product Design	CA_1
	Customer Satisfaction	CA_4
	Customer Response	CA_4
	Product Life Cycle	CL_4
	On-Time Delivery	CA_3
	Price Stability	CA_4
	Certificates	CL_1
Internal Business Processes	Creating Communication between Processes	$CL_1 + CA_4$
	Focus on Business (Core Abilities)	$CL_1 + CA_4$
	Adequate and Flexible Processes	CA_1
	Use of Information Technology	$CL_1 + CA_4$
	Change in Business Processes	$CL_1 + CA_4$
	The Nature of Management	$CL_1 + CA_4$
	Organizational Structure	$CL_1 + CA_4$
	Production Methodology	$CL_1 + CA_4$
	Value-Added Processes	CL_3
	Production Planning	CL_1
	Available Processes	CL_1
	Type of Automation	$CL_1 + CA_4$
	Technical Reasons	$CL_1 + CA_4$
	Internal Control	CL_4

Fig. 6.6 Defined indicators in the product design and development process by circular lean-agility

	Major Changes..	CL_4
	Risk Sharing..	CL_4
	Outsourcing..	$CL_1 + CA_4$
	Production...	$CL_1 + CA_4$
	Efficiency..	$CL_1 + CA_4$
	Quality...	$CL_1 + CA_4$
Organizational Capacity	Increasing the Performance Level of Suppliers and their Culture......	CA_4
	Competence and Talent of Employees...................................	CA_2
	Delegation of Management Authority....................................	$CL_1 + CA_4$
	Creating Motivation and Culture..	CA_4
	Acquiring Skills and Expertise..	CA_2
	Research and Development...	$CL_1 + CA_4$
	Knowledge Management...	$CL_1 + CA_4$
	Employee Satisfaction...	CL_1
	Design Improvement..	$CL_1 + CA_4$
	Time Management...	$CL_1 + CA_4$
	Expert Employees...	CA_2
	Empowerment...	$CL_1 + CA_2 + CA_4$
	Innovation...	$CL_1 + CA_4$

Fig. 6.6 (continued)

In the next step, we will assess the indicators required in a circular supply chain management system or supply chain network. Figure 6.7 shows the indicators can thoroughly cover the following two main components as follows:

1. The components of a balanced scorecard model
2. The components of circular lean-agility

This approach can guide the readers or practitioners in the assessment of the whole performance of the mentioned circular supply chain management. Figure 6.7 shows the required information that is a vital source of our approach in the presented case study in this chapter. Pay attention to the interaction between factors.

To prepare Fuzzy DEMATEL Matrix, the required abbreviations are needed as follows (Table 6.1):

Tables 6.2, 6.3, 6.4, and Fig. 6.8 (Z-MR Chart) show the extracted results related to the application of Fuzzy DEMATEL and Fuzzy ANP in the product design and development process as follows:

All data extracted from the selected fuzzy method in this domain (product design and development process) regarding the considered target are under control.

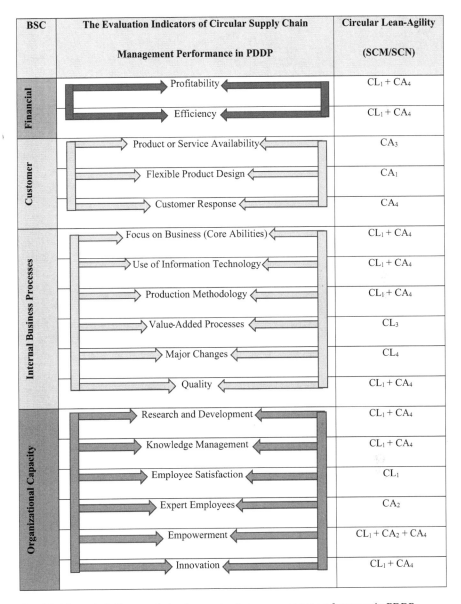

BSC	The Evaluation Indicators of Circular Supply Chain Management Performance in PDDP	Circular Lean-Agility (SCM/SCN)
Financial	Profitability	$CL_1 + CA_4$
	Efficiency	$CL_1 + CA_4$
Customer	Product or Service Availability	CA_3
	Flexible Product Design	CA_1
	Customer Response	CA_4
Internal Business Processes	Focus on Business (Core Abilities)	$CL_1 + CA_4$
	Use of Information Technology	$CL_1 + CA_4$
	Production Methodology	$CL_1 + CA_4$
	Value-Added Processes	CL_3
	Major Changes	CL_4
	Quality	$CL_1 + CA_4$
Organizational Capacity	Research and Development	$CL_1 + CA_4$
	Knowledge Management	$CL_1 + CA_4$
	Employee Satisfaction	CL_1
	Expert Employees	CA_2
	Empowerment	$CL_1 + CA_2 + CA_4$
	Innovation	$CL_1 + CA_4$

Fig. 6.7 The main indicators of circular supply chain management performance in PDDP

6.6 Conclusion

In today's competitive world, only organizations can survive and compete that are constantly evaluating and improving their performance and have an acceptable level of performance at the right time, according to the standards of the day. In this

Table 6.1 The required abbreviations in Fuzzy DEMATEL

F_{11}	Profitability
F_{12}	Efficiency
C_{11}	Product or Service Availability
C_{12}	Flexible Product Design
C_{13}	Customer Response
IBP_{11}	Focus on Business (Core Abilities)
IBP_{12}	Use of Information Technology
IBP_{13}	Production Methodology
IBP_{14}	Value-Added Processes
IBP_{15}	Major Changes
IBP_{16}	Quality
OC_{11}	Research and Development
OC_{12}	Knowledge Management
OC_{13}	Employee Satisfaction
OC_{14}	Expert Employees
OC_{15}	Empowerment
OC_{16}	Innovation

chapter, based on the model presented in the book for circular supply chain networks or circular supply chain management using fuzzy techniques and benefiting from quality engineering techniques, we applied the concepts of circular lean agility in a product design and development system as a case study. The vital goal of this chapter was to show our readers, who are probably involved in supply chain or production design, how they can use the presented model in the design and development of a product, for example by prioritizing all required actions that they should do. We will not forget that the basic philosophy of using fuzzy logic is to reflect uncertainty in calculations. By using the model presented in this book and receiving the opinion of practitioners (such as experts) of the organization, it was determined that the prioritization of the measures needed in the product design and development unit (case study) for the effective performance of the circular supply chain management system by combining fuzzy concepts, quality engineering techniques (DOE +SPC), and circular lean-agility. To conclude, the final goal of the chapter is to prioritize the important indicators of the product design and development process to plan for the establishment of circular lean agility thinking. The necessity of carrying out this study (a case study in this chapter) is rooted in the fact that planners, to select and prioritize plans at the time of planning and also to monitor the progress of plans at the time of implementation, need criteria and metrics to help them evaluate the conditions of the product design and development process. Table 6.5 shows the required actions (prioritization) as follows:

To achieve maximum efficiency and stability in the performance of a circular supply chain management system in product design and development, using fuzzy techniques, quality engineering techniques, and the concepts of circular lean agility, we recommend priorities 1–17 to production designers or the relevant practitioners including experts or managers in the different levels of a design department.

Table 6.2 The Fuzzy DEMATEL Matrix (1)

	F_{11}	F_{12}	C_{11}	C_{12}	C_{13}	IBP_{11}	IBP_{12}	IBP_{13}	IBP_{14}	IBP_{15}	IBP_{16}	OC_{11}	OC_{12}	OC_{13}	OC_{14}	OC_{15}	OC_{16}
F_{11}	–	0	2	1	0	2	1	1	1	1	1	1	3	2	1	1	0
F_{12}	4	–	0	2	0	2	1	1	1	1	1	2	3	4	2	2	3
C_{11}	4	4	–	0	0	4	0	0	0	0	0	1	3	3	0	1	0
C_{12}	3	4	3	–	0	4	2	3	3	3	3	0	3	3	1	0	3
C_{13}	4	3	4	0	–	0	1	0	1	0	3	3	3	0	0	1	0
IBP_{11}	4	4	3	2	3	–	0	3	3	3	2	2	3	2	2	3	3
IBP_{12}	3	3	2	2	3	3	–	0	3	3	2	2	3	2	2	3	2
IBP_{13}	3	3	3	0	3	2	2	–	0	2	0	1	1	1	0	0	1
IBP_{14}	3	3	2	1	2	3	3	2	–	0	0	2	1	0	0	0	2
IBP_{15}	3	3	3	2	3	3	2	1	1	–	0	2	2	2	3	0	3
IBP_{16}	4	4	0	3	3	2	0	2	0	0	–	0	3	1	0	1	1
OC_{11}	3	3	0	0	3	0	0	3	0	2	0	–	0	0	0	0	0
OC_{12}	3	3	1	2	2	1	1	2	0	1	0	3	–	0	3	2	4
OC_{13}	3	3	0	0	1	0	1	0	0	0	2	2	0	–	0	4	4
OC_{14}	3	3	2	1	3	1	0	1	1	0	2	3	0	0	–	0	4
OC_{15}	3	3	0	0	2	2	0	0	2	2	0	0	2	0	2	–	3
OC_{16}	3	3	1	0	3	1	0	2	3	3	0	3	3	2	4	4	–
	53	49	26	16	31	30	14	21	19	21	16	27	33	22	20	22	33

Table 6.3 The Fuzzy DEMATEL Matrix (2)

	F_{11}	F_{12}	C_{11}	C_{12}	C_{13}	IBP_{11}	IBP_{12}	IBP_{13}	IBP_{14}	IBP_{15}	IBP_{16}	OC_{11}	OC_{12}	OC_{13}	OC_{14}	OC_{15}	OC_{16}	
F_{11}	–	0.00%	7.69%	6.25%	0.00%	6.67%	7.14%	4.76%	5.26%	4.76%	6.25%	3.70%	9.09%	9.09%	5.00%	4.55%	0.00%	80.22%
F_{12}	7.55%	–	0.00%	12.50%	0.00%	6.67%	7.14%	4.76%	5.26%	4.76%	6.25%	7.41%	9.09%	18.18%	10.00%	9.09%	9.09%	117.76%
C_{11}	7.55%	8.16%	–	0.00%	0.00%	13.33%	0.00%	0.00%	0.00%	0.00%	0.00%	3.70%	9.09%	13.64%	0.00%	4.55%	0.00%	60.02%
C_{12}	5.66%	8.16%	11.54%	–	0.00%	13.33%	14.29%	14.29%	15.79%	14.29%	18.75%	0.00%	9.09%	13.64%	5.00%	0.00%	9.09%	152.91%
C_{13}	7.55%	6.12%	15.38%	0.00%	–	0.00%	7.14%	0.00%	5.26%	0.00%	18.75%	11.11%	9.09%	0.00%	0.00%	4.55%	0.00%	84.96%
IBP_{11}	7.55%	8.16%	11.54%	12.50%	9.58%	–	0.00%	14.29%	15.79%	14.29%	12.50%	7.41%	9.09%	9.09%	10.00%	13.64%	9.09%	164.60%
IBP_{12}	5.66%	6.12%	7.69%	12.50%	9.58%	10.00%	–	14.29%	15.79%	14.29%	12.50%	7.41%	9.09%	9.09%	10.00%	13.64%	6.06%	149.51%
IBP_{13}	5.66%	6.12%	11.54%	0.00%	9.58%	6.67%	14.29%	–	0.00%	9.52%	0.00%	3.70%	3.03%	4.55%	0.00%	0.00%	3.03%	77.78%
IBP_{14}	5.66%	6.12%	7.69%	6.25%	6.45%	10.00%	21.43%	9.52%	–	0.00%	0.00%	7.41%	3.03%	0.00%	0.00%	0.00%	6.06%	89.63%
IBP_{15}	5.66%	6.12%	11.54%	12.50%	9.68%	10.00%	14.29%	4.76%	5.26%	–	0.00%	7.41%	6.06%	9.09%	15.00%	0.00%	9.09%	126.46%
IBP_{16}	7.55%	8.16%	0.00%	18.75%	9.68%	6.67%	0.00%	9.52%	0.00%	0.00%	–	0.00%	9.09%	4.55%	0.00%	4.55%	3.03%	81.54%
OC_{11}	5.66%	6.12%	0.00%	0.00%	9.68%	0.00%	0.00%	9.52%	0.00%	9.52%	0.00%	–	0.00%	0.00%	0.00%	0.00%	0.00%	45.27%
OC_{12}	5.66%	6.12%	3.85%	12.50%	6.45%	3.33%	7.14%	9.52%	0.00%	4.76%	0.00%	11.11%	–	0.00%	15.00%	9.09%	12.12%	106.67%
OC_{13}	5.66%	6.12%	0.00%	0.00%	3.23%	0.00%	7.14%	0.00%	0.00%	0.00%	12.50%	7.41%	0.00%	–	0.00%	0.00%	12.12%	72.36%
OC_{14}	5.66%	6.12%	7.69%	6.25%	9.68%	3.33%	0.00%	4.76%	5.26%	0.00%	12.50%	11.11%	6.06%	0.00%	–	18.18%	12.12%	84.49%
OC_{15}	5.66%	6.12%	0.00%	0.00%	6.45%	6.67%	0.00%	0.00%	10.53%	9.52%	0.00%	0.00%	6.06%	0.00%	10.00%	–	9.09%	70.10%
OC_{16}	5.66%	6.12%	3.85%	0.00%	9.68%	3.33%	0.00%	9.52%	15.79%	14.29%	0.00%	11.11%	9.09%	9.09%	20.00%	18.18%	–	135.71%
	100.00%	100.00%	100.00%	100.00%	100.00%	100.00%	100.00%	100.00%	100.00%	100.00%	100.00%	100.00%	100.00%	100.00%	100.00%	100.00%	100.00%	

Table 6.4 The [Fuzzy DEMATEL + Fuzzy ANP] Matrix

Name	FDEMATEL + FANP	
	Target	Total priority
F_{11}	1.00	0.80
F_{12}	1.00	1.18
C_{11}	1.00	0.60
C_{12}	1.00	1.53
C_{13}	1.00	0.85
IBP_{11}	1.00	1.65
IBP_{12}	1.00	1.50
IBP_{13}	1.00	0.78
IBP_{14}	1.00	0.90
IBP_{15}	1.00	1.26
IBP_{16}	1.00	0.82
OC_{11}	1.00	0.45
OC_{12}	1.00	1.07
OC_{13}	1.00	0.72
OC_{14}	1.00	0.84
OC_{15}	1.00	0.70
OC_{16}	1.00	1.36

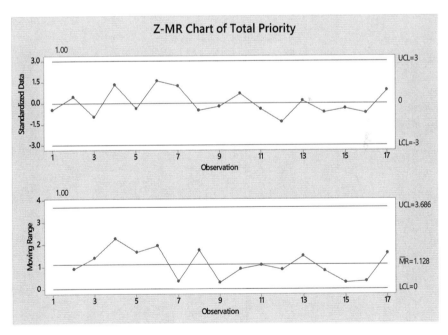

Fig. 6.8 The data demonstration in the domain of product design and development process (PDDP)

Table 6.5 The prioritization of required actions in the product design and development process (PDDP)

Prioritization number	Factor title	Factor value
Priority Number 1	Focus on Business (Core Abilities)	1.65
Priority Number 2	Attention to Flexible Product Design	1.53
Priority Number 3	Use of Information Technology	1.50
Priority Number 4	Attention to Innovation	1.36
Priority Number 5	Attention to Major Changes	1.26
Priority Number 6	Attention to Efficiency	1.18
Priority Number 7	Using Knowledge Management	1.07
Priority Number 8	Using Value-Added Processes	0.90
Priority Number 9	Commitment to Customer Response	0.85
Priority Number 10	Using Expert Employees	0.84
Priority Number 11	Attention to Quality	0.82
Priority Number 12	Attention to Profitability	0.80
Priority Number 13	Using Production Methodology	0.78
Priority Number 14	Achieving Employee Satisfaction	0.73
Priority Number 15	Attention to Empowerment	0.70
Priority Number 16	Product or Service Availability	0.60
Priority Number 17	Doing Research and Development	0.45

Exercises

1. There is information about four fuzzy techniques in the proposed indices for a circular SCM as follows:

The indices in circular SCM	Fuzzy methods				Y_{io}
	Fuzzy method number 1	Fuzzy method number 2	Fuzzy method number 3	Fuzzy method number 4	
I_1	65.25	46.51	37.22	58.57	
I_2	45.22	36.21	57.26	28.33	
I_3	65.24	56.41	67.57	78.52	
I_4	84.83	76.82	67.59	88.21	
I_5	25.83	56.37	77.56	90.33	
I_6	26.62	66.25	57.52	77.82	
I_7	55.25	55.51	66.26	58.62	
I_8	54.21	35.21	56.94	87.73	
I_9	94.38	55.82	86.89	27.82	
Y_{oj}					

Perform a complete analysis of variance for fuzzy techniques and the related indicators in circular supply chain management. Explain the results of the analysis in detail.

2. How can we combine the circular thinking elements into the supply chain management elements? What is the best approach to do it?
3. What are the best independent variables in the measurement of circular supply chain management performance? How can we categorize them?
4. How can we define a general target function in a supply chain network? How can we define a comprehensive restriction for the relevant supply chain network by the circular thinking elements?
5. What are the best differential equations to cover the overall productivity and sustainability level achieved by circular supply chain management influenced by the circular thinking elements?

© The Author(s), under exclusive license to Springer Nature Singapore Pte Ltd. 2024
R. Rostamkhani, T. Ramayah, *Navigating Circular Supply Chains*, SpringerBriefs in Operations Management, https://doi.org/10.1007/978-981-97-4704-7

6. Is it possible to rewrite the matrices of pairwise comparisons of circular thinking elements for supply chain network elements according to the four indicators of the balanced scorecard model as follows:

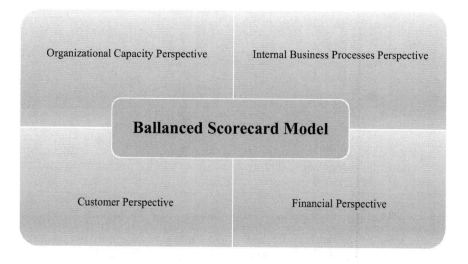

7. Could you explain some independent variables related to economic, environmental, and social issues for measuring the performance of a circular supply chain network in detail?

8. What is the most profound effect of applying the elements of circular thinking in the elements of supply chain management? How can we prove it? How can we show it?

9. Can the approach of this book be suitable for implementing a similar project in the circular economy?)The combination of circular thinking in the elements of the economy)

10. What is the best way to prove and show the added value of using elements of circular thinking in the elements of supply chain management or economy?

Appendix

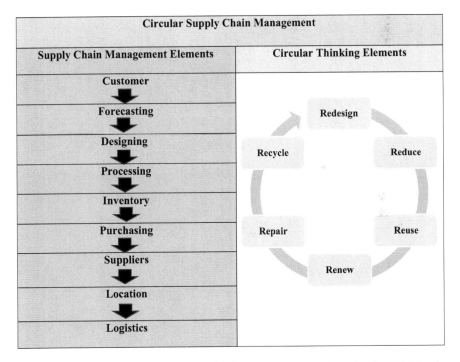

This appendix includes all the extracted information on applying circular thinking in the supply chain management elements using fuzzy techniques (FDEMATEL + FANP). Please pay attention to the different information on each component under the influence of circular thinking and the relevant model's priorities.

A.1 Customers

	Redesign	Reduce	Reuse	Renew	Repair	Recycle
Redesign	-	1	3	4	0	2
Reduce	0	-	3	4	1	1
Reuse	2	2	-	3	0	2
Renew	3	3	2	-	1	1
Repair	1	2	1	3	-	2
Recycle	6	10	11	16	2	8

	Redesign	Reduce	Reuse	Renew	Repair	Recycle
Redesign	-	0.20	0.60	0.80	0.00	0.40
Reduce	0.00	-	0.60	0.80	0.20	0.20
Reuse	0.40	0.40	-	0.60	0.00	0.40
Renew	0.60	0.60	0.40	-	0.20	0.20
Repair	0.20	0.40	0.20	0.60	-	0.40
Recycle	0.00	0.40	0.40	0.40	0.00	-

	Redesign	Reduce	Reuse	Renew	Repair	Recycle	
Redesign	-	10.0%	27.3%	25.0%	0.0%	25.0%	87.3%
Reduce	0.0%	-	27.3%	25.0%	50.0%	12.5%	114.8%
Reuse	33.3%	20.0%	-	18.8%	0.0%	25.0%	97.1%
Renew	50.0%	30.0%	18.2%	-	50.0%	12.5%	160.7%
Repair	16.7%	20.0%	9.1%	18.8%	-	25.0%	89.5%
Recycle	0.0%	20.0%	18.2%	12.5%	0.0%	-	50.7%
	100.0%	100.0%	100.0%	100.0%	100.0%	100.0%	

Redesign	Redesigning the original needs to the customer's demands
Reduce	Reducing extra time and cost to fulfill the customer's demands
Reuse	Reusing all previous sources in meeting the customer's demands
Renew	Renewing all required sources in meeting the customer's demands
Repair	Repairing all physical assets required for the different aspects of demands
Recycle	Recycling all previous experience or products in meeting the customer's demands

Customers

Element of CT	Priority
Renew	1
Reduce	2
Reuse	3

Target	Total Priority
1.00	0.87
1.00	1.15
1.00	0.97
1.00	1.61
1.00	0.90
1.00	0.51

A.2 Forecasting

Redesign	Redesigning the original needs to the demand predicting
Reduce	Reducing extra time and cost to fulfill the demand predicting
Reuse	Reusing all previous sources in meeting the demand predicting
Renew	Renewing all required sources in meeting the demand predicting
Repair	Repairing all physical assets required for the different aspects of predicting
Recycle	Recycling all previous experience or products in meeting the demand predicting

Forecasting

Element of CT	Priority
Reuse	1
Redesign	2
Renew	3

Target	Total Priority
1.00	1.23
1.00	0.73
1.00	1.29
1.00	1.13
1.00	0.81
1.00	0.81

	Redesign	Reduce	Reuse	Renew	Repair	Recycle
Redesign	-	1	4	4	1	4
Reduce	0	-	4	4	0	3
Reuse	2	2	-	3	1	4
Renew	3	3	2	-	0	3
Repair	1	2	1	3	-	3
Recycle	0	2	2	2	1	-
	6	10	13	16	3	17

	Redesign	Reduce	Reuse	Renew	Repair	Recycle
Redesign	-	0.20	0.80	0.80	0.20	0.80
Reduce	0.00	-	0.80	0.80	0.00	0.60
Reuse	0.40	0.60	-	0.60	0.20	0.80
Renew	0.60	0.60	0.40	-	0.00	0.60
Repair	0.20	0.40	0.20	0.60	-	0.60
Recycle	0.00	0.40	0.40	0.40	0.20	-

	Redesign	Reduce	Reuse	Renew	Repair	Recycle	
Redesign	-	10.0%	30.8%	25.0%	33.3%	23.5%	122.6%
Reduce	0.0%	-	30.8%	25.0%	0.0%	17.6%	73.4%
Reuse	33.3%	20.0%	-	18.8%	33.3%	23.5%	128.9%
Renew	50.0%	30.0%	15.4%	-	0.0%	17.6%	113.0%
Repair	16.7%	20.0%	7.7%	12.5%	-	17.6%	80.8%
Recycle	0.0%	20.0%	15.4%	12.5%	33.3%	-	81.2%
	100.0%	100.0%	100.0%	100.0%	100.0%	100.0%	100.0%

A.3 Designing

	Description
Redesign	Redesigning the original needs to the specifications
Reduce	Reducing extra time and cost to fulfill the specifications
Reuse	Reusing all previous sources in meeting the specifications
Renew	Renewing all required sources in meeting the specifications
Repair	Repairing all physical assets required for the different specifications
Recycle	Recycling all previous experience or products in meeting the specifications

	Redesign	Reduce	Reuse	Renew	Repair	Recycle
Redesign	-	4	4	4	0	3
Reduce	4	-	3	2	1	1
Reuse	4	2	-	3	0	2
Renew	3	3	2	-	1	1
Repair	3	0	1	0	-	1
Recycle	14	11	12	11	2	8

	Redesign	Reduce	Reuse	Renew	Repair	Recycle
Redesign	-	0.80	0.80	0.80	0.00	0.60
Reduce	0.80	-	0.60	0.40	0.20	0.20
Reuse	0.80	0.40	-	0.60	0.00	0.40
Renew	0.60	0.60	0.40	-	0.20	0.20
Repair	0.60	0.00	0.20	0.00	-	0.20
Recycle	0.00	0.40	0.40	0.40	0.00	-

	Redesign	Reduce	Reuse	Renew	Repair	Recycle	
Redesign	-	36.4%	33.3%	36.4%	0.0%	37.5%	143.6%
Reduce	28.6%	-	25.0%	18.2%	50.0%	12.5%	134.3%
Reuse	28.6%	18.2%	-	27.3%	0.0%	25.0%	99.0%
Renew	21.4%	27.3%	16.7%	-	50.0%	12.5%	127.9%
Repair	21.4%	0.0%	8.3%	0.0%	-	12.5%	42.3%
Recycle	0.0%	18.2%	16.7%	18.2%	0.0%	-	53.0%
	100.0%	100.0%	100.0%	100.0%	100.0%	100.0%	100.0%

Designing

Element of CT	Priority
Redesign	1
Reduce	2
Renew	3

Target	Total Priority
1.00	1.44
1.00	1.34
1.00	0.99
1.00	1.28
1.00	0.42
1.00	0.53

A.4 Processing

	Description
Redesign	Redesigning the original needs to the quality control
Reduce	Reducing extra time and cost to fulfill the quality control
Reuse	Reusing all previous sources in meeting the quality control
Renew	Renewing all required sources in meeting the quality control
Repair	Repairing all physical assets required for the different aspects of control
Recycle	Recycling all previous experience or products in meeting the quality control

Processing

Element of CT	Priority
Renew	1
Reduce	2
Recycle	3

Target	Total Priority
1.00	0.42
1.00	1.21
1.00	0.97
1.00	1.25
1.00	0.96
1.00	1.20

	Redesign	Reduce	Reuse	Renew	Repair	Recycle
Redesign	-	1	1	1	1	1
Reduce	1	-	3	3	4	4
Reuse	1	2	-	3	4	2
Renew	3	3	3	-	3	3
Repair	2	2	2	3	-	2
Recycle	3	2	3	4	2	-
	10	10	11	14	14	12

	Redesign	Reduce	Reuse	Renew	Repair	Recycle
Redesign	-	0.20	0.20	0.20	0.20	0.20
Reduce	0.20	-	0.60	0.60	0.80	0.80
Reuse	0.20	0.40	-	0.60	0.80	0.40
Renew	0.60	0.60	0.40	-	0.60	0.60
Repair	0.40	0.40	0.40	0.60	-	0.40
Recycle	0.60	0.40	0.60	0.80	0.40	-

	Redesign	Reduce	Reuse	Renew	Repair	Recycle	
Redesign	-	10.0%	9.1%	7.1%	7.1%	8.3%	41.7%
Reduce	10.0%	-	27.3%	21.4%	28.6%	33.3%	120.6%
Reuse	10.0%	20.0%	-	18.2%	28.6%	16.7%	96.7%
Renew	30.0%	30.0%	18.2%	-	21.4%	25.0%	124.6%
Repair	20.0%	20.0%	18.2%	21.4%	-	16.7%	96.3%
Recycle	30.0%	20.0%	27.3%	28.6%	14.3%	-	120.1%
	100.0%	100.0%	100.0%	100.0%	100.0%	100.0%	100.0%

A.5 Inventory

Redesign	Redesigning the original needs to the inventory management
Reduce	Reducing extra time and cost to fulfill the inventory management
Reuse	Reusing all previous sources in meeting the inventory management
Renew	Renewing all required sources in meeting the inventory management
Repair	Repairing all physical assets required for the different assessment of inventory
Recycle	Recycling all previous experience or products in meeting the inventory management

Inventory

Element of CT	Priority
Renew	1
Recycle	2
Reuse	3

Target	Total Priority
1.00	0.27
1.00	0.55
1.00	1.42
1.00	1.59
1.00	0.61
1.00	1.56

	Redesign	Reduce	Reuse	Renew	Repair	Recycle
Redesign	-	1	0	1	0	1
Reduce	0	-	1	2	2	1
Reuse	2	2	-	4	4	4
Renew	4	4	4	-	3	3
Repair	2	2	1	1	-	1
Recycle	4	4	4	3	3	-
	12	13	10	11	12	10

	Redesign	Reduce	Reuse	Renew	Repair	Recycle
Redesign	-	0.20	0.00	0.20	0.00	0.20
Reduce	0.00	-	0.40	0.40	0.40	0.20
Reuse	0.40	0.40	-	0.80	0.80	0.80
Renew	0.80	0.80	0.80	-	0.60	0.60
Repair	0.40	0.40	0.20	0.20	-	0.20
Recycle	0.80	0.80	0.80	0.60	0.60	-

	Redesign	Reduce	Reuse	Renew	Repair	Recycle	
Redesign	-	7.7%	0.0%	9.1%	0.0%	10.0%	26.8%
Reduce	0.0%	-	10.0%	18.2%	16.7%	10.0%	54.8%
Reuse	16.7%	15.4%	-	36.4%	33.3%	40.0%	141.7%
Renew	33.3%	30.8%	40.0%	-	25.0%	30.0%	159.1%
Repair	16.7%	15.4%	10.0%	9.1%	-	10.0%	61.1%
Recycle	33.3%	30.8%	40.0%	27.3%	25.0%	-	156.4%
	100.0%	100.0%	100.0%	100.0%	100.0%	100.0%	

A.6 Purchasing

Purchasing	
Element of CT	**Priority**
Reuse	1
Recycle	2
Reduce	3

Target	**Total Priority**
1.00	0.35
1.00	1.19
1.00	1.56
1.00	1.10
1.00	0.44
1.00	1.36

Redesign	Redesigning the original needs to the suppliers evaluation
Reduce	Reducing extra time and cost to fulfill the suppliers evaluation
Reuse	Reusing all previous sources in meeting the suppliers evaluation
Renew	Renewing all required sources in meeting the suppliers evaluation
Repair	Repairing all physical assets required for the different aspects of purchasing
Recycle	Recycling all previous experience or products in meeting the suppliers evaluation

	Redesign	Reduce	Reuse	Renew	Repair	Recycle
Redesign	-	1	0	2	1	0
Reduce	0	-	3	3	3	3
Reuse	2	2	-	4	4	4
Renew	2	2	2	-	2	2
Repair	1	1	1	1	-	0
Recycle	3	3	2	2	3	-
	8	9	8	12	13	9

	Redesign	Reduce	Reuse	Renew	Repair	Recycle
Redesign	-	0.20	0.00	0.40	0.20	0.00
Reduce	0.00	-	0.60	0.60	0.60	0.60
Reuse	0.40	0.40	-	0.80	0.80	0.80
Renew	0.40	0.40	0.40	-	0.40	0.40
Repair	0.20	0.20	0.20	0.20	-	0.00
Recycle	0.60	0.60	0.40	0.40	0.60	-

	Redesign	Reduce	Reuse	Renew	Repair	Recycle	
Redesign	-	11.1%	0.0%	16.7%	7.7%	0.0%	35.5%
Reduce	0.0%	-	37.5%	25.0%	23.1%	33.3%	118.9%
Reuse	25.0%	22.2%	-	33.3%	30.8%	44.4%	155.8%
Renew	25.0%	22.2%	25.0%	-	15.4%	22.2%	109.8%
Repair	12.5%	11.1%	12.5%	8.3%	-	0.0%	44.4%
Recycle	37.5%	33.3%	25.0%	16.7%	23.1%	-	135.6%
	100.0%	100.0%	100.0%	100.0%	100.0%	100.0%	100.0%

A.7 Suppliers

	Redesign	Reduce	Reuse	Renew	Repair	Recycle
Redesign	-	1	0	2	0	1
Reduce	0	-	3	2	3	2
Reuse	1	1	-	1	1	1
Renew	3	2	1	-	1	1
Repair	0	2	2	3	-	2
Recycle	5	9	8	10	5	7

	Redesign	Reduce	Reuse	Renew	Repair	Recycle
Redesign	-	0.20	0.00	0.40	0.00	0.20
Reduce	0.00	-	0.60	0.40	0.60	0.40
Reuse	0.20	0.20	-	0.20	0.20	0.20
Renew	0.60	0.60	0.40	-	0.20	0.20
Repair	0.20	0.40	0.20	0.60	-	0.40
Recycle	0.00	0.40	0.40	0.40	0.00	-

	Redesign	Reduce	Reuse	Renew	Repair	Recycle	Total
Redesign	-	11.1%	0.0%	20.0%	0.0%	14.3%	45.4%
Reduce	0.0%	-	37.5%	20.0%	60.0%	28.6%	146.1%
Reuse	20.0%	11.1%	-	10.0%	20.0%	14.3%	75.4%
Renew	60.0%	33.3%	25.0%	-	20.0%	14.3%	152.6%
Repair	20.0%	22.2%	12.5%	30.0%	-	28.6%	113.3%
Recycle	0.0%	22.2%	25.0%	20.0%	0.0%	-	67.2%
	100.0%	100.0%	100.0%	100.0%	100.0%	100.0%	

	Definition
Redesign	Redesigning the original needs to the supplier's quality
Reduce	Reducing extra time and cost to fulfill the supplier's quality
Reuse	Reusing all previous sources in meeting the supplier's quality
Renew	Renewing all required sources in meeting the supplier's quality
Repair	Repairing all physical assets required for the different aspects of suppliers
Recycle	Recycling all previous experience or products in meeting the supplier's quality

Suppliers

Element of CT	Priority
Renew	1
Reduce	2
Repair	3

Target	Total Priority
1.00	0.45
1.00	1.46
1.00	0.75
1.00	1.53
1.00	1.13
1.00	0.67

A.8 Location

	Location	
Element of CT	**Priority**	
Repair	1	
Reuse	2	
Renew	3	

Target	Total Priority
1.00	0.75
1.00	0.91
1.00	1.25
1.00	0.92
1.00	1.36
1.00	0.81

Redesign	Redesigning the original needs to the location determination	
Reduce	Reducing extra time and cost to fulfill the location determination	
Reuse	Reusing all previous sources in meeting the location determination	
Renew	Renewing all required sources in meeting the location determination	
Repair	Repairing all physical assets required for the different aspects of location	
Recycle	Recycling all previous experience or products in meeting the location determination	

	Redesign	Reduce	Reuse	Renew	Repair	Recycle
Redesign	-	2	2	2	2	2
Reduce	0	-	3	3	3	3
Reuse	4	3	-	4	3	3
Renew	4	3	2	-	3	2
Repair	4	4	4	3	-	3
Recycle	3	2	1	15	2	-
	13	14	12	15	13	13

	Redesign	Reduce	Reuse	Renew	Repair	Recycle
Redesign	-	0.40	0.40	0.40	0.40	0.40
Reduce	0.00	-	0.60	0.60	0.60	0.60
Reuse	0.80	0.60	-	0.80	0.60	0.60
Renew	0.40	0.60	0.40	-	0.60	0.40
Repair	0.80	0.80	0.80	0.60	-	0.40
Recycle	0.60	0.40	0.20	0.60	0.60	-

	Redesign	Reduce	Reuse	Renew	Repair	Recycle	
Redesign	-	14.3%	16.7%	13.3%	15.4%	15.4%	75.1%
Reduce	0.0%	-	25.0%	20.0%	23.1%	23.1%	91.2%
Reuse	30.8%	21.4%	-	26.7%	23.1%	23.1%	125.0%
Renew	15.4%	28.6%	16.7%	-	23.1%	15.4%	91.9%
Repair	30.8%	14.3%	33.3%	20.0%	-	23.1%	135.8%
Recycle	23.1%	21.4%	8.3%	20.0%	15.4%	-	81.1%
	100.0%	100.0%	100.0%	100.0%	100.0%	100.0%	100.0%

A.9 Logistics

Logistics

Element of CT	Priority
Repair	1
Reuse	2
Redesign	3

Redesign	Redesigning the original needs to the movement and storage
Reduce	Reducing extra time and cost to fulfill the movement and storage
Reuse	Reusing all previous sources in meeting the movement and storage
Renew	Renewing all required sources in meeting the movement and storage
Repair	Repairing all physical assets required for the different aspects of M &S
Recycle	Recycling all previous experience or products in meeting the movement and storage

	Redesign	Reduce	Reuse	Renew	Repair	Recycle
Redesign	-	1	2	2	3	3
Reduce	0	-	2	2	2	2
Reuse	3	3	-	3	3	3
Renew	2	2	2	-	1	1
Repair	4	4	4	4	-	4
Recycle	0	1	1	2	2	-
	9	11	11	13	11	13

	Redesign	Reduce	Reuse	Renew	Repair	Recycle
Redesign	-	0.20	0.40	0.40	0.60	0.60
Reduce	0.00	-	0.40	0.40	0.40	0.40
Reuse	0.60	0.60	-	0.60	0.60	0.60
Renew	0.40	0.40	0.40	-	0.20	0.20
Repair	0.80	0.80	0.80	0.80	-	0.80
Recycle	0.00	0.20	0.20	0.40	0.40	-

	Redesign	Reduce	Reuse	Renew	Repair	Recycle	
Redesign	-	9.1%	18.2%	15.4%	27.3%	23.1%	93.0%
Reduce	0.0%	-	18.2%	15.4%	18.2%	15.4%	67.1%
Reuse	33.3%	27.3%	-	23.1%	27.3%	23.1%	134.0%
Renew	22.2%	18.2%	18.2%	-	30.8%	7.7%	75.4%
Repair	44.4%	36.4%	36.4%	30.8%	-	30.8%	178.7%
Recycle	0.0%	9.1%	9.1%	15.4%	18.2%	-	51.7%
	100.0%	100.0%	100.0%	100.0%	100.0%	100.0%	

Target	Total Priority
1.00	0.93
1.00	0.67
1.00	1.34
1.00	0.75
1.00	1.79
1.00	0.52